Humanity's Next Frontier

SPACE COLONIES

Charting Our Cosmic Legacy and the
Evolution of Humanity

OSMAN KARAKAS

About Book

Book Title: Space Colonies: *Humanity's Next Frontier*

Subtitle: Charting Our Cosmic Legacy and the Evolution of Humanity

Type: Digital E-Book

Format: Word/PDF

Size: 6X9 inches - 15.24X22.89 cm

Total Pages: **405**

E-mail: okarakas@hotmail.com

Web: www.osmankarakas.com

CONTENTS

Preface:

In the vast tapestry of the cosmos, humanity has found its next frontier: space colonies. As we stand on the precipice of a new era, one marked by interstellar aspirations and extraterrestrial habitats, this book seeks to illuminate the profound journey upon which we are embarking.

Space Colonies: Humanity's Next Frontier is the culmination of extensive research, tireless exploration, and boundless curiosity. Its pages unfold like the chapters of a cosmic epic, detailing the remarkable feats of human engineering, the ingenuity of scientific discovery, and the resilience of the human spirit against the backdrop of space.

Within these chapters, you will traverse the unimaginable distances between stars and planets, exploring the intricacies of self-sustaining habitats and the complexities of life beyond Earth's atmosphere. But this book is more than a mere account of technological marvels and scientific achievements; it delves deep into the human aspects of space colonization. It explores how these celestial outposts challenge our perceptions of identity, culture, and ethics, raising profound questions about our place in the universe.

As an award-winning journalist, editor, and researcher, my passion for unraveling the mysteries of space and human nature has driven this endeavor. I have sought to capture the essence of space colonies not just as scientific endeavors but as crucibles of human potential, where creativity, collaboration, and adaptability are tested and refined.

For educators and students, this book offers a comprehensive resource, delving into the practicalities of space colonization while igniting the imagination with the possibilities it presents. For the curious minds and avid readers, it provides a fascinating journey through the unexplored realms of space, painting vivid pictures of life in extraterrestrial habitats.

I extend my heartfelt gratitude to the readers whose thirst for knowledge propels the pursuit of cosmic understanding. Together, we embark on a voyage that challenges the boundaries of what humanity can achieve, inviting you to peer through the windows of our future homes among the stars.

As you turn these pages, may you find inspiration, curiosity, and a sense of wonder. May you be propelled toward the limitless horizons of the universe, where our dreams and aspirations take flight, carried by the inexorable spirit of exploration.

Welcome, dear reader, to a celestial journey that begins on these pages and stretches to the furthest reaches of the cosmos.

Warm regards,

Osman Karakas

Introduction: Embarking on a Cosmic Odyssey

In the limitless reaches of space, a new chapter in the story of humanity is being written—one that transcends the boundaries of Earth and propels us into the cosmic expanse. This book, *Space Colonies: Humanity's Next Frontier*, invites you on a captivating odyssey, delving deep into the heart of space exploration and the remarkable habitats that humans have built beyond our home planet.

With each passing moment, our understanding of the universe expands, revealing not only the vastness of space but also the potential for human endeavor in the cosmos. This exploration is not just a scientific feat but a testament to human ingenuity, resilience, and unwavering curiosity. Within these pages, you will embark on a journey that takes you beyond the stars, exploring the intricacies of space colonies, their impact on our society, and the myriad possibilities they hold for our collective future.

The narrative unfolds against the backdrop of distant galaxies and uncharted celestial realms, where pioneers and visionaries have dared to dream of a future among the stars. Through meticulous research, captivating storytelling, and thought-provoking analysis, this book endeavors to unravel the mysteries of space colonization, offering insights into the challenges faced, the innovations developed, and the profound implications for humanity.

We will venture into the heart of these cosmic habitats, exploring the intricate ecosystems, groundbreaking technologies, and the human spirit that thrives amidst the cosmic challenges. From the vast domes of self-sustaining habitats to the laboratories conducting groundbreaking research, we will witness the unfolding saga of human adaptation and transformation in the face of the unknown.

Beyond the scientific marvels, we will delve into the societal and ethical dimensions of space colonization. How does living among the stars shape our perspectives on identity, culture, and cooperation? What ethical considerations guide our interactions in the cosmic neighborhood? These questions and more will be explored, inviting you to contemplate the broader implications of our cosmic ambitions.

This book is not merely a collection of facts and figures; it is a chronicle of human ambition and the enduring spirit that propels us toward the cosmos. It is an exploration of the limitless potential within us, urging us to reach for the stars and embrace the challenges and wonders that await us there.

As you turn the pages of this book, you are not just reading about space colonies; you are embarking on a cosmic odyssey, experiencing the triumphs and tribulations of humanity's boldest endeavor. So, fasten your seatbelts, for the journey to the stars begins now, and the adventure promises to be nothing short of extraordinary.

Welcome to *Space Colonies: Humanity's Next Frontier*. Let the cosmic odyssey commence.

Chapter 1: The Vision of Space Colonies

Early Dreams and Astronomical Pioneers

In the tapestry of human history, there existed dreamers—early astronomers who cast their gaze upon the stars and envisioned a destiny beyond Earth. From the ancient astronomers of Babylon who tracked the movements of celestial bodies to the Renaissance polymaths who dared to challenge Earth's central position in the universe, these pioneers sowed the seeds of cosmic ambition.

Their calculations and observations, often made with rudimentary instruments, revealed patterns in the night sky. Ptolemy's geocentric model, the prevailing theory for centuries, depicted Earth as the center of the universe, around which planets and stars orbited. However, Copernicus dared to challenge this notion, proposing a heliocentric model that placed the Sun at the center, setting the stage for a revolution in astronomical thought.

The Space Age Dawns

The mid-20th century witnessed the birth of the Space Age—a period defined by humanity's audacious leap beyond Earth's atmosphere. Yuri Gagarin, a Soviet cosmonaut, became the first human to orbit the Earth in 1961, marking a historic moment in the annals of space exploration. His mission aboard Vostok 1 captured the world's imagination, illustrating the tangible reality of space travel.

Gagarin's achievement, however, was not an isolated event but rather the beginning of an era marked by bold endeavors. The United States, in response, embarked on the Apollo program. In 1969, Neil Armstrong and Buzz Aldrin set foot on the Moon, their words echoing through the vast lunar expanse, "That's one small step for [a] man, one giant leap for mankind." The Apollo 11 mission showcased humanity's ability to overcome gravitational confines and venture into the cosmic unknown.

Visionaries and Their Ambitions

As the Space Age unfolded, a new breed of visionaries emerged—individuals whose ambitions transcended the confines of our home planet. Elon Musk, the founder of SpaceX, became synonymous with Mars colonization. His audacious vision included establishing a self-sustaining human colony on the Red Planet, a concept that had long been relegated to the realm of science fiction.

SpaceX, under Musk's leadership, achieved unprecedented milestones. The development of reusable rocket technology transformed the economics of space travel. Falcon 9's successful landings, both on Earth and on autonomous drone ships, heralded a new era of space exploration where the barriers to entry were lowered, opening avenues for further innovation.

This vision was not limited to Mars alone. Musk's endeavors encompassed the establishment of a lunar base and, ultimately, the prospect of interplanetary travel. His bold pronouncements and tangible progress rekindled humanity's fascination with the cosmos, propelling space exploration into a new era of possibility.

Societal Motivations and Benefits

Space colonization represents more than a scientific venture—it embodies humanity's quest for survival and progress. In an age marked by environmental challenges and finite resources, space colonies offer a beacon of hope. They serve as sanctuaries preserving Earth's biodiversity and cultural heritage, ensuring the continuity of our species.

Moreover, the societal implications of space colonization are profound. These colonies become melting pots of human diversity, fostering cultural exchange and understanding. The challenges faced in extraterrestrial environments necessitate cooperation, transcending geopolitical boundaries. The shared pursuit of cosmic frontiers unites humanity in a common endeavor, highlighting the potential for peaceful coexistence.

Challenges and Triumphs

However, the path to space colonization is fraught with challenges. Technological hurdles, ethical dilemmas, and the harsh realities of space present formidable obstacles. This chapter delves deeply into the inherent struggles faced by humanity— the moments of setbacks and failures that tested the resolve of scientists, engineers, and explorers. It explores the intricacies of problem-solving, innovation, and the sheer determination that propelled humanity forward despite the adversities.

Triumphs Over Adversity: Celebrating Human Ingenuity

Amidst these challenges, humanity showcases remarkable resilience and creativity. This section of the chapter illuminates the triumphs over adversity, celebrating the moments of groundbreaking success, innovation, and the unwavering spirit of exploration. From overcoming the limitations of technology to finding ingenious solutions to life support systems, each triumph stands as a testament to human ingenuity.

The Evolution of Spacecraft: A Journey Through Innovation

One of the remarkable triumphs in space exploration lies in the evolution of spacecraft. From the rudimentary capsules of the early Space Age to the sophisticated vessels that navigate the cosmos today, engineers and scientists have pushed the boundaries of technological innovation.

Apollo's Lunar Module, a marvel of engineering, demonstrated humankind's ability to land on another celestial body and return safely to Earth. This feat, achieved in the vacuum of space and the harsh lunar environment, represented a triumph of meticulous planning, precision engineering, and sheer audacity.

In the following decades, space agencies and private enterprises alike embraced innovation. Space shuttles, like NASA's iconic Space Shuttle program, became symbols of reusable space travel. These shuttles ferried astronauts to and from Earth's orbit, paving the way for the construction of the International Space Station (ISS). The ISS stands as a testament to international collaboration, a floating laboratory where scientists conduct experiments, test technologies, and live in the microgravity of space.

Life Support Systems: Nurturing Humanity Beyond Earth

Surviving in the harsh vacuum of space demands ingenious life support systems. Overcoming the challenge of providing astronauts with air, water, and sustenance in a closed-loop environment required innovative solutions.

The recycling of water became a crucial triumph. Advanced filtration systems purify astronauts' urine and condensation, transforming them into potable water. Closed-loop systems, where oxygen is continuously regenerated and carbon dioxide removed, mimic Earth's natural processes, ensuring astronauts have a breathable atmosphere.

Innovations in hydroponics and aeroponics enabled the cultivation of crops in space, providing astronauts with fresh produce. These self-contained ecosystems not only nourish the crew but also serve as essential components in future space colonies, where sustainable agriculture will be the key to long-term habitation.

Human Adaptation: Overcoming Biological Challenges

The human body, finely tuned to Earth's conditions, faces numerous challenges in space. Microgravity affects bone density and muscle mass, posing health risks to astronauts. Yet, humanity's resilience shines through in the face of adversity.

Researchers have developed exercise regimens and specialized equipment to counteract the effects of prolonged weightlessness. Astronauts engage in daily physical activities, utilizing advanced exercise devices that simulate resistance, ensuring their muscles and bones remain strong.

Biomedical research delves into the intricacies of human biology, seeking solutions to challenges posed by space environments. Studies on the International Space Station, where astronauts serve as test subjects, provide valuable data on the human body's response to extended spaceflight, guiding future space colonization efforts.

Innovative Energy Solutions: Powering the Cosmic Journey

In the vast expanse of space, energy is life. Space colonies demand a consistent and renewable energy source to power life support systems, scientific experiments, and human activities. Triumphs in energy solutions have reshaped the way humans harness power beyond Earth's atmosphere.

Solar panels, capable of capturing sunlight and converting it into electricity, have become integral components of spacecraft and space stations. Advancements in solar technology, including lightweight and efficient panels, ensure a steady supply of energy even in the distant reaches of our solar system.

Nuclear power offers a potent energy source for space colonies. Radioisotope thermoelectric generators (RTGs) harness the heat produced by radioactive decay to generate electricity. RTGs have powered numerous deep-space missions, such as the Mars rovers, providing vital energy in environments where solar panels might be inefficient.

Entrepreneurial Innovation: Opening New Frontiers

The 21st century witnessed a paradigm shift in space exploration. Private enterprises, driven by visionary entrepreneurs, entered the cosmic arena. SpaceX, Blue Origin, and other companies embarked on ventures that redefined the space industry landscape.

Reusable rocket technology, pioneered by SpaceX, drastically reduced launch costs and increased the frequency of space missions. The Falcon 9's vertical landings, once deemed science fiction, became routine occurrences, paving the way for ambitious projects like Mars colonization.

Private enterprises also turned their attention to lunar exploration. Lunar landers and rovers, funded by private initiatives, explored the Moon's surface, mapping its terrain and identifying resources. These ventures set the stage for future lunar bases, serving as waypoints for further exploration into the cosmos.

The Spirit of Innovation: Nurturing Curiosity and Exploration

Triumphs over adversity are not solely confined to technological achievements; they reside in the spirit of exploration that permeates the cosmos. Curiosity, the driving force behind humanity's cosmic odyssey, fuels our insatiable quest for knowledge.

From the intricate dance of celestial bodies to the mysteries of black holes, each discovery expands the boundaries of human understanding. Telescopes, both ground-based and spaceborne, capture the light from distant stars and galaxies, revealing the universe's wonders in unprecedented detail.

In the search for extraterrestrial life, scientists explore the most extreme environments on Earth, such as deep-sea hydrothermal vents and acidic hot springs. These studies provide insights into the potential habitats for life beyond our planet, guiding the search for alien organisms in our solar system and beyond.

Conclusion of Triumphs Over Adversity

The journey into space, marked by triumphs over adversity, exemplifies humanity's ability to overcome challenges through innovation, determination, and collaboration. These triumphs stand as beacons of inspiration, illuminating the path toward space colonization. As the cosmic narrative unfolds, these triumphs herald a future where humanity's indomitable spirit propels us further into the infinite expanse of the universe.

Chapter 2: Technological Innovations

The Essence of Spacefaring Technology: Propulsion Systems

Central to the endeavor of space colonization is the propulsion technology that propels spacecraft beyond Earth's gravitational embrace. Traditional chemical propulsion, employing the energetic reactions of rocket propellants, served as the pioneering force that lifted humanity into space. Yet, the quest for more efficient and faster propulsion methods spurred a wave of innovation.

Ion propulsion systems, harnessing the power of ionized gases accelerated by electromagnetic fields, offer exceptional efficiency and endurance. These engines enable spacecraft to reach previously unattainable velocities, transforming interplanetary travel and reducing transit times between celestial bodies.

Nuclear thermal propulsion represents another frontier in space propulsion. By utilizing nuclear reactors to heat propellants, these systems generate high thrust, making them ideal for rapid interplanetary travel. Nuclear thermal rockets promise to revolutionize missions to Mars and beyond, unlocking the outer reaches of our solar system for exploration.

Life Support Technologies: Sustaining Human Life in Space

Ensuring the sustenance of human life in the unforgiving environment of space requires sophisticated life support technologies. Closed-loop systems, where air and water are continuously recycled, play a pivotal role in providing astronauts with the essentials for survival.

Advanced filtration technologies purify the air within spacecraft, removing carbon dioxide and trace contaminants. Water recovery systems meticulously extract moisture from the air, condensate, and even urine, converting them into clean drinking water. These innovations form the backbone of self-sustaining habitats, essential for long-term space colonization efforts.

Bridging Humanity and Machines: Robotics in Space Exploration

The integration of robotics into space exploration has ushered in a new era of exploration and discovery. Robotic systems, equipped with advanced sensors and manipulators, undertake tasks deemed hazardous or impractical for human astronauts. These machines venture into uncharted territories, paving the way for human colonization.

Robotic rovers, exemplified by NASA's Mars rovers, traverse alien landscapes, conducting geological surveys and searching for signs of past life. These autonomous explorers relay invaluable data back to Earth, guiding the selection of landing sites for future manned missions.

In orbital construction, robotic arms, such as those on the International Space Station, assemble modules and conduct intricate repairs. These machines, operated by astronauts or ground-based controllers, extend the capabilities of spacefarers, enabling the assembly of colossal structures in the microgravity of space.

AI and Space Exploration: The Rise of Intelligent Explorers

Artificial intelligence (AI) has emerged as a cornerstone of space exploration, revolutionizing mission planning, data analysis, and autonomous decision-making. AI-driven systems process vast datasets, identifying patterns and anomalies that might elude human observers. These insights inform scientific research and enhance mission safety.

Machine learning algorithms enable autonomous navigation, guiding spacecraft through intricate trajectories and complex orbital maneuvers. AI-powered robotic systems adapt to unforeseen challenges, adjusting their actions in real-time, ensuring mission success in the face of uncertainty.

Harvesting the Cosmic Energy: Sustainable Power Sources

In the boundless expanse of space, energy is the lifeblood of exploration. Sustainable energy sources ensure the perpetual operation of space habitats, scientific instruments, and life support systems. Solar power, harnessed through photovoltaic panels, captures the energy of starlight, providing a consistent and renewable source of electricity.

Beyond our solar system, space colonies might utilize space-based solar arrays, capturing the unfiltered energy of distant stars. These vast arrays would beam energy wirelessly to receiving stations, powering habitats in the deepest reaches of interstellar space. Such innovative solutions herald a future where energy constraints do not limit the cosmic aspirations of humanity.

The Nexus of Innovation: Transforming Space Colonization

The technological innovations explored in this chapter represent the vanguard of space exploration. Propulsion systems capable of interstellar travel, life support technologies ensuring self-sufficiency, intelligent robotic explorers expanding the boundaries of knowledge, and sustainable energy sources powering the cosmic odyssey — these advancements converge to transform space colonization from a dream into a tangible reality.

As humanity embarks on missions to the Moon, Mars, and beyond, these innovations will pave the way for permanent human presence in space. The synergy of these technologies propels humanity toward a future where the stars themselves are within reach, where the cosmos becomes not just a frontier but a new home.

Chapter 3: Challenges of Interplanetary Travel

Understanding Cosmic Radiation: The Silent Foe of Space Travel

The interstellar void is not empty; it is filled with cosmic radiation, a silent foe lurking in the cosmic shadows. These high-energy particles, originating from distant stars and supernovae, pose a significant threat to astronauts venturing beyond Earth's protective atmosphere. Understanding this radiation is paramount; shielding technologies must evolve, ensuring that human explorers are shielded from this invisible danger. Collaborative efforts between physicists, engineers, and biologists delve into the mysteries of cosmic radiation, seeking innovations that will safeguard the pioneers of space colonization.

The Toll on Human Physiology: Long-Duration Spaceflights and Beyond

Long-duration spaceflights exact a toll on the human body, challenging the very essence of human physiology. Microgravity, the pervasive condition of space, weakens muscles and bones, demanding innovative exercise regimens to maintain the physical strength of astronauts. Fluid shifts within the body, altering vision and impairing cardiovascular function, highlight the intricate interplay between humans and their cosmic environment. The biological clock of astronauts, entrained by Earth's diurnal cycle, is disrupted, posing challenges to sleep patterns and mental health. Understanding these physiological transformations is essential, paving the way for tailored interventions and medical innovations that will enable humanity to endure the rigors of space exploration.

The Mind in Isolation: Psychological Challenges Beyond Earth

Isolation, confinement, and the unyielding vastness of space weigh heavily on the human psyche. Psychological challenges, intricately woven into the fabric of space exploration, demand nuanced approaches. Astronauts, isolated for extended durations, grapple with a myriad of emotions — from exhilaration to homesickness. Communication delays with Earth, a stark reality of interplanetary travel, intensify the sense of isolation, necessitating resilient mental fortitude. Scientists delve into the depths of psychology, exploring coping mechanisms and cultivating the mental resilience required for voyages that span the cosmic expanse.

Closed-Loop Life Support Systems: A Symphony of Sustainability

In the microcosm of space habitats, life is a delicate balance, sustained by closed-loop life support systems — intricate symphonies of sustainability. Every breath exhaled, every drop of water consumed, every morsel of food ingested is part of this orchestrated dance. Air is meticulously filtered, ensuring a continuous supply of oxygen while removing carbon dioxide. Water, recycled with unparalleled efficiency, flows through purifiers and filters, emerging as pristine hydration. Waste, once discarded, is transformed into vital resources, fostering the growth of plants and sustaining the ecosystem. The development of closed-loop systems is not merely a technical achievement; it is a testament to humanity's capacity for ingenuity, mirroring the cyclical rhythms of nature within the confines of space.

Mastering Artificial Gravity: The Key to Long-Term Human Habitation

In the absence of Earth's gravitational embrace, the human body languishes. Muscles atrophy, bones weaken, and the essence of human physiology falters. Artificial gravity, a concept as old as science fiction itself, holds the key to mitigating these detrimental effects. Centrifugal force, harnessed within rotating space habitats, mimics Earth's gravity, offering a solution that could enable long-term human habitation in space. The engineering marvels required to create these rotating habitats challenge the very limits of human knowledge. Innovations in propulsion, materials science, and structural integrity converge in the quest to master artificial gravity, ensuring that future space colonies become cradles of human civilization amidst the cosmic wilderness.

The Odyssey Continues: Navigating the Cosmic Challenges

The challenges of interplanetary travel, woven into the fabric of space exploration, stand as milestones on humanity's cosmic odyssey. Cosmic radiation, the silent nemesis, demands shielding innovations; physiological transformations necessitate tailored medical interventions. Psychological fortitude, the cornerstone of mental health in space, requires ongoing research and support systems. Closed-loop life support systems, the embodiment of sustainability, pave the way for self-sufficient habitats. Artificial gravity, the linchpin of human well-being, offers the promise of long-term habitation beyond Earth. The odyssey continues, forging ahead with determination and unyielding curiosity, as humanity ventures into the cosmic unknown.

Chapter 4: Building Habitats in Space

Pioneering Space Architecture: Where Science and Creativity Converge

In the boundless canvas of space, architects and engineers embark on a pioneering journey, shaping the future of human habitation beyond Earth. Space architecture, a fusion of science and creativity, envisions habitats that defy terrestrial constraints. The symbiosis of form and function becomes paramount as architects explore innovative designs optimized for the unique challenges posed by celestial bodies. From the harsh regolith of the Moon to the enigmatic terrain of Mars, these architects navigate the complexities of extraterrestrial landscapes, ensuring that the habitats they envision become sanctuaries for human exploration and innovation.

Blueprints for Alien Worlds: Designing for Mars, the Red Frontier

Mars, the red frontier, captivates the imagination of architects and engineers, beckoning them to create habitats that harmonize with its rugged landscapes. Redefining the concept of 'home,' these Martian habitats become oasis-like structures, designed to shield inhabitants from the planet's thin atmosphere and dust storms. Architects delve into the Martian regolith, contemplating the use of indigenous materials to construct habitats that seamlessly integrate with the Martian environment. 3D printing technology, a beacon of innovation, emerges as a cornerstone in this architectural odyssey, allowing for the creation of intricate structures layer by layer, transforming Martian dust into habitable spaces that echo the resilience of human ingenuity.

Lunar Havens: Crafting Habitats Amidst Moon Dust

The Moon, Earth's celestial companion, becomes a crucible of lunar architecture, challenging architects to craft habitats amidst moon dust. Lunar regolith, a powdery terrain rich in potential resources, becomes the building blocks for lunar havens. Architects envision habitats with domed structures, cocooned beneath a protective layer, shielding inhabitants from the Moon's harsh climate. The lunar surface, once a barren expanse, transforms into a hub of human activity, as architects and engineers collaborate to create habitats that endure the lunar night and embrace the lunar day. The hum of 3D printers echoes through the lunar landscape, as habitats materialize, ushering in a new era of lunar exploration and habitation.

Beyond Celestial Bodies: Space Habitats as Microcosms of Earth

Beyond the confines of individual celestial bodies, space architects envision habitats as microcosms of Earth, nurturing life amidst the cosmic void. These habitats, reminiscent of self-sustaining ecosystems, become sanctuaries where humans coexist with nature in harmony. Within these space habitats, plants thrive under artificial sunlight, purifying the air and providing sustenance. Aquatic ecosystems, enclosed within transparent chambers, echo the gentle rhythms of Earth's oceans, fostering biodiversity and sustaining life. Architects weave a tapestry of life within these habitats, creating verdant oases amidst the vastness of space. The vision of these space colonies, flourishing with life, stands as a testament to humanity's aspiration to transform the cosmic wilderness into a cradle of life.

Innovations in 3D Printing: Sculpting the Future of Space Habitats

Central to the realization of extraterrestrial habitats is the groundbreaking technology of 3D printing, a cornerstone of innovation in space architecture. Architects and engineers collaborate with robotic assistants, crafting habitats layer by layer, intricately sculpting structures that withstand the cosmic challenges. The marriage of advanced materials and robotic precision gives rise to habitats that are not merely shelters but testaments to human resilience. Moon dust and Martian regolith, once deemed inhospitable, become the raw materials for these architectural marvels. As 3D printers hum with activity, space habitats emerge, redefining the cosmic landscape and heralding a new era where human civilization extends its reach beyond the stars.

The Ethereal Beauty of Space Habitats: Where Art and Science Converge

Within the confines of space habitats, art and science converge, giving rise to ethereal beauty amidst the cosmic expanse. Artists and architects collaborate, transforming the sterile walls of space habitats into canvases of creativity. Luminescent artworks adorn the corridors, casting a soft glow reminiscent of starlight. Gardens bloom with vibrant colors, cultivated under carefully calibrated lighting, adding a touch of nature's brilliance to the extraterrestrial environment. Architects design habitats with panoramic windows, offering breathtaking views of nearby celestial bodies, connecting inhabitants with the cosmic wonders beyond. In these artistic sanctuaries, the human spirit soars, finding solace in the boundless beauty of the cosmos.

Harmonizing with Extraterrestrial Environments: Lessons from Nature

In the pursuit of space habitats, architects and scientists draw inspiration from the wisdom of nature, harmonizing human creations with extraterrestrial environments. Biomimicry becomes a guiding principle, as architects emulate the ingenious designs found in nature. From the resilient exoskeletons of insects to the self-healing properties of plants, these lessons from Earth's biodiversity inspire innovations in habitat design. Materials that mimic the strength of spider silk and the durability of seashells find their way into the construction of space habitats. Nature, the ultimate architect, becomes a mentor, guiding humanity toward sustainable habitats that endure the rigors of space, echoing the resilient harmony found within Earth's ecosystems.

Inhabitants of the Cosmic Havens: Adapting to Extraterrestrial Life

Within these cosmic havens, inhabitants adapt to the rhythms of extraterrestrial life, embracing a lifestyle intricately woven into the fabric of space habitats. Daily routines synchronize with the artificial cycles of light and darkness, mirroring the cosmic dance of nearby celestial bodies. Inhabitants become stewards of closed-loop ecosystems, tending to the plants that purify their air and recycle their water. Workshops hum with activity, as residents engage in scientific research, artistic pursuits, and the cultivation of sustenance. Communities thrive, forging bonds amidst the cosmic wilderness, creating a new way of life that transcends terrestrial norms. Adaptation becomes the essence of survival, as humanity transforms from inhabitants of Earth to pioneers of the cosmic frontier.

The Future Unveiled: Space Habitats as Cradles of Innovation

In the tapestry of human innovation, space habitats emerge as cradles of creativity, nurturing a future where human civilization extends its reach to the stars. These habitats become incubators of technological marvels, inspiring innovations that benefit life both within and beyond Earth. Advancements in recycling technologies, perfected within closed-loop ecosystems, find applications in terrestrial cities, addressing the pressing challenges of environmental sustainability. Insights gained from extraterrestrial agriculture revolutionize farming practices, ensuring food security for a burgeoning global population. The fusion of art and science within space habitats permeates terrestrial cultures, enriching humanity's creative expression. The future, unveiled within these cosmic havens, heralds a new era where the boundless potential of human ingenuity knows no celestial bounds.

Chapter 5: Social Dynamics in Space Colonies

Isolation and Unity: Navigating the Cosmic Social Landscape

Within the confined quarters of space colonies, a profound social experiment unfolds, challenging the very essence of human interaction. Isolation, both breathtaking and daunting, becomes the backdrop against which social dynamics play out. In this cosmic crucible, inhabitants embark on a journey of self-discovery, forging bonds amidst the vastness of space. The unity of purpose becomes a guiding beacon, illuminating the path toward collective resilience. From the silent corridors of lunar habitats to the bustling hubs of Martian colonies, the tapestry of social structures takes shape, weaving a narrative of unity, camaraderie, and the human spirit's indomitable resolve.

Confinement and Connection: The Paradox of Cosmic Proximity

Confinement within space colonies presents a paradoxical challenge: the closeness of inhabitants juxtaposed against the vast cosmic distances that surround them. In this paradox, connection becomes both a lifeline and a source of tension. Inhabitants navigate the intricacies of personal space, adapting to the nuances of communal living. Shared resources, communal dining, and collaborative workspaces become the norm, fostering a sense of interdependence. Yet, within the confines of limited space, conflicts can arise, testing the patience and understanding of spacefarers. The delicate balance between personal privacy and collective harmony becomes a focal point, shaping the social fabric of space colonies.

Impact of the Cosmic Environment: Shaping Human Behavior

The unique environment of space, with its microgravity and cosmic radiation, exerts a profound influence on human behavior. Spacefarers, adapted to the terrestrial embrace of Earth, find themselves in an alien setting, challenging the very core of their existence. The absence of gravity transforms daily activities, from simple tasks like eating and sleeping to the intricacies of scientific experiments. Adaptation becomes key, as inhabitants learn to maneuver in this novel environment, embracing innovative solutions and technologies. The cosmic environment becomes a silent teacher, shaping the behavior of spacefarers, instilling resilience, and fostering a spirit of exploration that transcends the confines of Earth.

Conflict and Resolution: Navigating the Cosmic Crossroads

Conflict, a natural facet of human interaction, finds its way even into the cosmic havens of space colonies. Differences in opinions, cultural backgrounds, and personal preferences can lead to tension within the confined quarters. Yet, within this crucible of conflict, the seeds of resolution are sown. Communication becomes a powerful tool, bridging the gaps between individuals and fostering understanding. Mediation and conflict resolution training become integral components of space colony life, equipping inhabitants with the skills necessary to navigate disagreements constructively. The cosmic crossroads become a training ground for emotional intelligence, where empathy and mutual respect become the pillars upon which harmonious relationships are built.

The Importance of Teamwork: The Foundation of Cosmic Societies

In the cosmic expanse, teamwork emerges as the cornerstone upon which the foundations of space colonies are laid. The synergy of diverse talents and expertise becomes a wellspring of innovation, propelling scientific research, engineering endeavors, and artistic pursuits to new heights. Inhabitants collaborate across disciplines, pooling their knowledge to overcome challenges and explore the mysteries of the universe. Teamwork fosters a sense of belonging, nurturing a collective identity that transcends national boundaries and cultural differences. The importance of teamwork reverberates through the corridors of space colonies, creating a tapestry of collaboration and mutual support that sustains the dreams of cosmic pioneers.

Psychological Support Systems: Nurturing the Mind in the Cosmic Wilderness

Amidst the awe-inspiring cosmic vistas, the human mind navigates its own odyssey, confronting the depths of isolation and the vastness of space. Psychological support systems become the bedrock upon which mental well-being is nurtured. Counselors and psychologists, trained in the nuances of space psychology, offer support to inhabitants, providing a listening ear amidst the cosmic silence. Virtual reality becomes a therapeutic tool, allowing spacefarers to reconnect with Earth, easing homesickness and offering moments of respite. Peer support networks, woven tightly within the social fabric of space colonies, become lifelines, offering companionship and understanding in the face of the cosmic unknown.

Unity in Diversity: Embracing Differences in the Cosmic Melting Pot

Within the cosmic melting pot of space colonies, diversity becomes a source of strength, a tapestry of cultures, languages, and traditions that enrich the social landscape. Inhabitants, hailing from different corners of Earth, bring with them a mosaic of perspectives, shaping the collective consciousness of cosmic societies. Cultural exchanges, celebrations of festivals, and shared culinary experiences become avenues through which unity in diversity is celebrated. Prejudices dissolve in the face of shared cosmic aspirations, paving the way for a future where the human family transcends the boundaries of Earth, embracing the vast cosmic tapestry as one unified civilization.

The Vision of Cosmic Harmony: Toward a Future of Unity and Exploration

In the heart of space colonies, amidst the challenges and triumphs of cosmic life, a vision of cosmic harmony takes shape. Inhabitants, bound by the shared dream of exploration, nurture a future where unity, understanding, and cooperation define human interactions. The cosmic wilderness, once a daunting expanse, becomes a canvas upon which humanity paints its aspirations. Cosmic pioneers, guided by the principles of unity and exploration, venture into the unknown, pushing the boundaries of human potential. In this vision of cosmic harmony, Earth's diverse cultures merge into a tapestry of cosmic civilization, fostering a future where the stars are not merely distant lights but beacons guiding humanity toward a destiny written among the galaxies.

Chapter 6: Extraterrestrial Resource Utilization

The Cosmic Harvest: Mining Asteroids and Moons

In the boundless reaches of space, asteroids and moons stand as celestial treasure troves, holding the key to sustaining life beyond Earth. Mining asteroids, with their rich deposits of metals, minerals, and rare elements, becomes a cornerstone of extraterrestrial resource utilization. These cosmic bodies, remnants of the early solar system, harbor minerals like platinum, gold, and rare earth elements in abundance. Space miners equipped with advanced robotic technologies delve into the ancient rocks, extracting invaluable resources that fuel innovation and economic growth within space colonies.

On the lunar frontier, Earth's faithful companion, the Moon, reveals its watery secrets. Frozen water, locked within lunar craters, becomes a lifeline for spacefarers. Lunar water extraction, a marvel of engineering, involves melting the frozen deposits, providing a precious source of drinking water and a vital component for life support systems. In the lunar twilight, specialized equipment transforms the barren lunar surface into an oasis of hydration, paving the way for sustained human presence on the Moon and beyond.

The Water of Life: Lunar Hydration and Beyond

Water, the elixir of life, takes center stage in the cosmic drama of colonization. Lunar water, extracted from the Moon's regolith, becomes the lifeblood of space colonies, supporting the hydration needs of inhabitants and serving as a crucial component in agricultural ventures. The vision of sustainable agriculture within space habitats becomes a reality, as lunar water irrigates fields of genetically modified crops, carefully nurtured under artificial light. In the cosmic dance of life, lunar water not only quenches the thirst of spacefarers but also irrigates the crops that sustain them, creating closed-loop ecosystems within space colonies.

Beyond Earth: The Potential of Extraterrestrial Resources

Extraterrestrial resources extend far beyond Earth's immediate vicinity, presenting tantalizing opportunities for the future of humanity. Asteroid mining ventures venture deeper into space, exploring the asteroid belt between Mars and Jupiter, where ancient celestial bodies, rich in precious metals, await their cosmic prospectors. Mining operations, conducted with precision and care, extract resources like nickel, cobalt, and iron, elements vital for constructing space habitats and spacecraft. These mining endeavors not only fuel the expansion of space colonies but also provide raw materials for the construction of interplanetary vessels, paving the way for the next giant leap in human exploration.

Fueling Exploration: The Role of Extraterrestrial Resources

Extraterrestrial resources serve as the lifeblood of cosmic exploration, propelling spacecraft deeper into the cosmos. Water extracted from the Moon becomes the raw material for propellant, powering spacecraft on their journeys to distant planets and celestial bodies. Hydrogen, extracted from lunar water through electrolysis, combines with oxygen to form powerful rocket fuel. In the cosmic void, where resources are scarce, the utilization of lunar water becomes a game-changer, reducing the cost of space travel and enabling missions of unprecedented scope and ambition.

Economic Frontiers: The Rise of Cosmic Industries

The utilization of extraterrestrial resources heralds the rise of cosmic industries, shaping the future of Earth's economy. Asteroid mining companies, once mere pioneers, evolve into powerful entities, driving economic growth and innovation. Precious metals and rare elements extracted from asteroids find their way back to Earth, becoming components of advanced technologies, luxury goods, and even works of art.

Lunar water becomes a valuable commodity, traded between space colonies and Earth, fueling the growth of interplanetary economies. The cosmic frontier, once an uncharted territory, becomes a bustling hub of economic activity, where the exchange of resources and knowledge paves the way for a prosperous future for humanity.

The Ethical Tapestry: Balancing Profit and Preservation

Amidst the economic opportunities of extraterrestrial resource utilization, ethical dilemmas loom large. The delicate balance between profit and preservation becomes a focal point of cosmic ethics. Environmental concerns arise as mining operations transform the celestial landscape, raising questions about the long-term impact on asteroid populations and lunar geology. Responsible mining practices, guided by international agreements and ethical frameworks, become essential. Spacefaring civilizations grapple with questions of ownership and stewardship, pondering the rightful custodianship of cosmic resources. Ethical debates echo through the halls of space colonies and Earth's governing bodies, shaping the moral compass of humanity's cosmic endeavors.

Cosmic Sustainability: A Vision for the Future

In the grand tapestry of the cosmos, sustainability emerges as the guiding principle of extraterrestrial resource utilization. Spacefarers, custodians of the celestial realm, embrace a vision where cosmic resources are harnessed responsibly, ensuring the preservation of celestial environments for future generations. Closed-loop systems within space colonies, utilizing extraterrestrial resources for life support and agriculture, become models of cosmic sustainability. Scientific research, conducted with reverence for celestial bodies, unveils the mysteries of the cosmos while respecting the delicate balance of cosmic ecosystems.

The Cosmic Mosaic: Humanity's Enduring Legacy

Extraterrestrial resource utilization, an odyssey that spans the vastness of space, becomes a chapter in humanity's enduring legacy. The cosmic mosaic, woven with threads of exploration, innovation, and ethical stewardship, paints a picture of a species capable of transcending its earthly confines. In the utilization of extraterrestrial resources, humanity finds not only economic opportunities but also a profound connection to the cosmos. The lessons learned in the cosmic crucible shape the future of Earth and its cosmic colonies, guiding the evolution of societies and civilizations. As spacefarers gaze upon distant stars, they carry with them the wisdom of cosmic sustainability, ensuring that the human legacy endures among the galaxies.

Chapter 7: Interplanetary Economics

The Cosmic Marketplace: Business Models in Space Colonization

In the vast expanse of the cosmos, new frontiers for business emerge. Space colonization becomes not only a scientific and exploratory endeavor but also a thriving marketplace. As humanity ventures into the stars, innovative business models take shape, fueled by the unique challenges and opportunities of space. Entrepreneurs and visionaries craft bold strategies, exploring ventures ranging from asteroid mining corporations to space tourism enterprises. The cosmic marketplace teems with potential, where traditional business paradigms blend with the unparalleled opportunities of the final frontier.

Investment Horizons: Funding the Cosmic Dream

Space exploration demands substantial investments, sparking a new era of financial creativity and collaboration. Venture capitalists, drawn by the promise of extraterrestrial resources and groundbreaking technologies, fund ambitious space startups. Government grants and international collaborations fuel scientific research and interplanetary missions. Crowdfunding platforms unite space enthusiasts worldwide, enabling grassroots support for space-related projects. The investment landscape, once bound by terrestrial boundaries, expands to encompass the boundless horizons of space, shaping the future of cosmic exploration.

Commercial Ventures: From Earth to the Stars

The commercialization of space activities propels humanity into a new age of cosmic enterprise. Private companies, ranging from aerospace giants to innovative startups, redefine the possibilities of space technology. Launch services become commonplace, with reusable rockets revolutionizing space travel. Satellites, essential for communication and Earth observation, populate the skies, providing services ranging from global internet coverage to weather forecasting. Lunar and Martian missions, once solely the realm of governments, now see private entities establishing research outposts and fuel depots, heralding an era where the stars are within reach of commercial ventures.

The Space Tourism Revolution: Holidays among the Stars

Space tourism emerges as a transformative industry, inviting adventurous souls to embark on celestial journeys. Space resorts, orbiting Earth or stationed on the Moon, offer unparalleled views of the cosmos. Suborbital flights become accessible to the public, allowing travelers to experience weightlessness and witness the curvature of their home planet. Lunar vacations become a reality, with tourists exploring the Moon's surface and marveling at Earth from the lunar horizon. Space enthusiasts, once dreamers of the stars, now become pioneers of space tourism, embracing the wonders of the cosmos firsthand.

The Ethical Quandaries: Balancing Profit and Ethics

The economic expansion into space raises profound ethical questions, challenging humanity's moral compass. Corporate interests collide with environmental preservation, as celestial bodies become sites for mining and resource extraction. Ethical frameworks are devised to ensure the responsible utilization of space resources, protecting celestial environments from overexploitation and pollution. Intellectual property rights and space patents provoke debates, questioning the ownership of cosmic knowledge and innovations. The delicate balance between profit and ethics becomes a central theme in interplanetary economics, guiding policymakers and entrepreneurs toward decisions that safeguard both economic interests and the cosmic heritage.

International Collaborations: The Global Economy of the Stars

Space exploration transcends national borders, fostering collaborations that shape the global economy of the stars. International partnerships become commonplace, with space agencies, research institutions, and private companies pooling resources and expertise. Joint missions to distant planets and celestial bodies become a testament to human collaboration, uniting nations in the pursuit of cosmic knowledge. Trade agreements and treaties govern the exchange of cosmic resources, ensuring equitable distribution and fostering peaceful coexistence among spacefaring nations. The global economy of the stars becomes a testament to humanity's ability to unite in the face of cosmic challenges, paving the way for a future where the stars belong to all.

The Entrepreneurial Spirit: Shaping the Cosmic Destiny

At the heart of interplanetary economics lies the entrepreneurial spirit, a driving force that propels humanity into the cosmic frontier. Visionaries and innovators, unbound by Earthly limitations, redefine the boundaries of what is possible. Space entrepreneurs pioneer new industries, from space manufacturing to extraterrestrial agriculture, transforming the cosmic landscape. The cosmic marketplace becomes a testament to human ingenuity, where startups and established enterprises collaborate, compete, and innovate, shaping the destiny of humanity among the stars.

Corporate Citizenship: A Cosmic Responsibility

As corporations extend their influence into space, a new paradigm of corporate citizenship emerges. Ethical practices, environmental stewardship, and social responsibility become integral to the corporate ethos. Space companies invest in sustainable technologies, ensuring that their ventures minimize their ecological impact on celestial bodies. Philanthropic initiatives support space education and research, empowering future generations with cosmic knowledge. Corporate citizenship becomes a cosmic responsibility, shaping the legacy of businesses among the stars and inspiring a new era of ethical entrepreneurship.

The Cosmic Renaissance: Economic Impact on Earth

The economic expansion into space reverberates across Earth, sparking a cosmic renaissance. Technological innovations born from space exploration find applications in terrestrial industries, revolutionizing healthcare, transportation, and communication. Earthbound economies flourish as investments in space technology yield advancements in renewable energy, artificial intelligence, and materials science. The pursuit of cosmic knowledge fuels curiosity and creativity, inspiring scientists, engineers, and entrepreneurs to push the boundaries of innovation. The cosmic renaissance becomes a testament to the interconnectedness of space and Earth, where advancements in the stars elevate the human experience on our home planet.

Conclusion: The Economic Odyssey of Space Colonization

As this chapter draws to a close, the economic odyssey of space colonization stands as a testament to humanity's entrepreneurial spirit and thirst for discovery. From the innovative business models shaping the cosmic marketplace to the ethical dilemmas guiding economic decisions, every aspect of interplanetary economics reflects the multifaceted nature of humanity's cosmic endeavor.

The stars, once distant and unreachable, now stand as beacons of economic opportunity and human potential. As entrepreneurs, investors, and dreamers gaze upward, they see not only the vastness of space but also the boundless possibilities awaiting those who dare to explore, innovate, and collaborate.

The economic odyssey of space colonization becomes a chapter in the epic saga of human exploration, where the pursuit of profit harmonizes with the quest for knowledge, propelling humanity toward a future where the stars are not just distant lights but thriving hubs of economic activity, creativity, and limitless potential.

Chapter 8: Legal and Ethical Considerations

The Cosmic Code: Legal Frameworks for Space Exploration

In the boundless expanse of space, the need for legal frameworks becomes paramount. Spacefaring nations come together to craft international treaties and agreements, shaping the legal landscape of cosmic exploration. The Outer Space Treaty, signed by numerous nations, establishes that celestial bodies are the province of all humankind, prohibiting the placement of nuclear weapons in space and limiting the use of the Moon and other celestial bodies exclusively for peaceful purposes.

The Moon Agreement further emphasizes that the Moon and its resources are the common heritage of humanity, fostering the responsible use and preservation of lunar resources for the benefit of all. Legal scholars and policymakers engage in discussions about property rights and resource extraction, paving the way for a legal framework that ensures equitable access to cosmic resources while safeguarding the celestial environments.

Property Rights in the Cosmos: Ownership and Stewardship

The question of property rights in space sparks debates that echo across terrestrial borders. Can celestial bodies be owned, or are they the shared heritage of humanity? Legal experts explore the concept of space property rights, considering scenarios where nations or private entities establish bases on the Moon or Mars. The notion of stewardship emerges, emphasizing responsible resource management and environmental preservation. Scholars delve into historical precedents, drawing parallels between space exploration and the Age of Exploration on Earth. The balance between private enterprise, national interests, and the common heritage of humanity becomes a focal point of legal discussions, guiding the formulation of laws that ensure fairness and sustainability in space colonization.

International Cooperation: Diplomacy among the Stars

Space exploration transcends geopolitical boundaries, necessitating international cooperation and diplomacy. Space agencies collaborate on joint missions, sharing scientific knowledge and technological expertise. International space stations become symbols of global collaboration, where astronauts from different nations work together in harmony. Legal frameworks for collaborative endeavors are established, addressing liability, intellectual property rights, and data sharing. Diplomats engage in dialogue, resolving disputes and fostering trust among spacefaring nations. International cooperation becomes a cornerstone of cosmic exploration, paving the way for peaceful coexistence and collaborative efforts in the cosmic frontier.

Preservation of Celestial Environments: Environmental Ethics in Space

As humanity ventures into space, ethical considerations extend beyond human interactions to encompass the preservation of celestial environments. Space ecologists study extraterrestrial ecosystems, exploring the potential for life on other planets and moons. Ethical guidelines are developed to ensure that human activities do not disrupt or harm indigenous extraterrestrial life forms, if they exist. Scientists and environmentalists advocate for the protection of pristine lunar landscapes and Martian terrains, recognizing their scientific and cultural value. Ethical codes of conduct are established for space explorers, emphasizing the responsible exploration and study of celestial bodies. The preservation of celestial environments becomes a testament to humanity's commitment to environmental ethics, reflecting a harmonious coexistence between human endeavors and the cosmic wilderness.

Conflict and Cooperation: Navigating Territorial Disputes

As space exploration advances, potential conflicts arise over territory and resources. Lunar and Martian regions rich in resources become focal points of international attention. Diplomats navigate territorial disputes, drawing on historical agreements such as the Antarctic Treaty as models for international collaboration in space. The establishment of buffer zones and shared research areas fosters cooperation while minimizing conflict. Ethical principles guide negotiations, ensuring that the preservation of celestial environments remains a priority. Spacefaring nations engage in dialogue, finding innovative solutions to disputes and setting precedents for responsible cosmic exploration. Conflict resolution becomes an art in the cosmic arena, where diplomacy and ethics pave the way for shared prosperity and scientific discovery.

Corporate Ventures and Cosmic Ethics: A Delicate Balance

Private corporations, driven by economic interests, enter the cosmic stage, raising ethical questions about their activities. Corporate ventures in space mining and resource extraction pose challenges to existing legal frameworks. Ethical entrepreneurs pioneer sustainable practices, emphasizing environmental responsibility and community engagement. Transparency and accountability become watchwords in the corporate cosmos, ensuring that business activities align with ethical principles. NGOs and advocacy groups monitor corporate behavior, promoting ethical standards and responsible conduct. The delicate balance between corporate interests, legal regulations, and ethical considerations becomes a focal point of debates, shaping the future of commercial ventures in space and setting standards for corporate citizenship beyond Earth.

The Future of Cosmic Ethics: Guiding Humanity's Destiny

In the cosmic odyssey, ethical considerations serve as the compass guiding humanity's destiny among the stars. Ethicists, scientists, diplomats, and entrepreneurs collaborate to establish ethical principles that safeguard celestial environments, preserve the common heritage of humanity, and ensure the equitable distribution of cosmic resources. The evolving nature of ethics in space exploration becomes a dynamic field, where new challenges spark innovative solutions and ethical frameworks adapt to the complexities of the cosmic frontier. As humanity's presence in space grows, ethical considerations become an integral part of cosmic decision-making, shaping the legacy of space exploration and inspiring future generations to explore the stars with wisdom, compassion, and a profound respect for the cosmic tapestry.

Conclusion: Navigating the Cosmic Ethical Landscape

As this chapter concludes, the exploration of legal and ethical considerations in space colonization stands as a testament to humanity's capacity for thoughtful, responsible exploration. The cosmic frontier, with its vastness and complexity, challenges humanity to transcend terrestrial conflicts and embrace a unified vision for the future. Legal frameworks and ethical principles become the foundation upon which space exploration rests, ensuring that the stars remain accessible to all and that the cosmic heritage of humanity is preserved for future generations. As the cosmic odyssey continues, legal experts, ethicists, and visionaries collaborate, forging a path where the exploration of space becomes a harmonious endeavor, uniting humanity in the pursuit of knowledge, cooperation, and a profound reverence for the cosmic wonders that await.

Chapter 9: Cultural Impact

The Cosmic Canvas: Space in Literature and Arts

Space, with its boundless mysteries and infinite possibilities, has been a wellspring of inspiration for artists and writers throughout history. Literature, in its various forms, paints the cosmic canvas with tales of interstellar adventures, alien encounters, and the human spirit's resilience amidst the stars. Visionary authors craft narratives that transcend the earthly realm, exploring themes of exploration, extraterrestrial life, and the evolution of human society in the cosmic tapestry. Poets, too, find solace in the cosmic expanse, weaving verses that echo the vastness of space and the timeless allure of distant celestial bodies. Artists, on canvas and screen, translate the wonders of the universe into visual symphonies, capturing the awe-inspiring beauty of nebulae, galaxies, and alien landscapes. Through brushstrokes and words, humanity ventures into the cosmic unknown, finding inspiration in the enigmatic depths of space.

Starlit Silver Screens: Space Exploration in Cinema

The silver screens of cinemas around the world flicker to life with visions of space exploration. Filmmakers, armed with imagination and cutting-edge technology, transport audiences to alien worlds, distant galaxies, and futuristic space colonies. Science fiction movies, from the golden age classics to contemporary blockbusters, delve into the realms of speculative science, showcasing advanced civilizations, time travel, and encounters with extraterrestrial beings. These cinematic odysseys ignite the human imagination, sparking discussions about the potential future of humanity in the cosmos. The melding of science and art in space-themed movies captivates viewers, bridging the gap between scientific curiosity and artistic expression. Directors and screenwriters, with an eye on scientific accuracy, craft narratives that blur the lines between fiction and reality, inviting audiences to contemplate the profound questions of existence and humanity's destiny among the stars.

The Creative Cosmos: Space in Visual Arts

Visual artists, with their diverse forms of expression, capture the essence of space exploration in their works. Painters, sculptors, and digital artists translate cosmic wonders into tangible art pieces, infusing their creations with a sense of wonder and reverence for the universe. Cosmic landscapes, alien civilizations, and futuristic space technologies find their place on gallery walls and exhibition spaces, inviting viewers to embark on imaginative journeys through the cosmos. The interplay of light and shadow, color and form, gives life to celestial bodies, creating a visual symphony that resonates with the human spirit's yearning for discovery. Space-themed art not only reflects the scientific advancements of the age but also serves as a catalyst for contemplation, encouraging viewers to ponder the mysteries of the universe and their place within it. Through artistic expression, humanity explores the infinite facets of space, transcending the confines of earthly existence and embracing the cosmic vastness.

Space in Music: Harmonies of the Celestial Spheres

The allure of space finds its resonance in music, where composers and musicians craft melodies inspired by the cosmic tapestry. Orchestral compositions evoke the grandeur of celestial bodies in motion, capturing the majesty of planets, stars, and galaxies through musical notes. Synthesizers and electronic music delve into the realm of the unknown, creating sonic landscapes that echo the enigmatic depths of space.

Space-themed musical compositions accompany space missions, providing a soundtrack to the exploration of the cosmos. From classical compositions that mirror the elegance of the cosmos to avant-garde pieces that challenge traditional musical boundaries, space-themed music transcends genres, inviting listeners on auditory journeys through the universe. The harmonies of the celestial spheres find their reflection in the melodies of human creation, creating a bridge between the cosmic wonders above and the creative spirit within.

The Impact on Human Imagination: Dreaming Beyond the Stars

Space exploration serves as a catalyst for human imagination, igniting dreams that transcend the boundaries of Earth. Children gaze at the night sky, inspired by the twinkling stars, envisioning adventures among distant planets and alien civilizations. Writers pen tales of cosmic exploration, architects design habitats on alien worlds, and scientists contemplate the potential for life beyond Earth. The collective imagination of humanity extends into the cosmos, fueled by the discoveries of space missions and the tantalizing possibility of extraterrestrial contact. Imagination becomes the driving force behind scientific innovation, artistic expression, and societal progress, propelling humanity toward a future where the boundaries of the known universe blur, and the unexplored territories of space beckon with endless possibilities.

Conclusion: The Cosmic Tapestry of Human Creativity

As this chapter concludes, the exploration of space's cultural impact reveals a profound truth:

humanity's creativity knows no bounds when inspired by the wonders of the universe. Through literature, cinema, visual arts, music, and the boundless realm of imagination, humanity weaves the cosmic tapestry, enriching the cultural heritage of our species. Space exploration becomes more than a scientific endeavor; it becomes a source of inspiration that transcends disciplines and connects people across continents.

The impact of space on culture mirrors humanity's innate curiosity, our endless quest for knowledge, and our ability to dream beyond the stars. As artists and thinkers continue to explore the cosmic unknown, the cultural legacy of space exploration evolves, shaping the narratives of future generations and inspiring the eternal pursuit of discovery, creativity, and the infinite wonders that await in the boundless expanse of space.

Chapter 10: Future Horizons

The Terraforming Odyssey: Crafting New Worlds

Terraforming, the monumental endeavor of transforming hostile extraterrestrial environments into habitable worlds, stands at the forefront of humanity's ambitions. In the pursuit of space colonization, scientists and engineers envision a future where barren landscapes are reshaped, atmospheres are revitalized, and ecosystems flourish. This section delves into the intricate processes of terraforming, exploring the scientific principles and technologies that could one day make alien worlds, such as Mars, suitable for human habitation. From the deployment of greenhouse gases to the cultivation of genetically modified organisms, humanity charts a course toward transforming desolate planets into thriving, Earth-like habitats. The challenges of terraforming, from the manipulation of planetary climates to the creation of sustainable ecosystems, are met with innovative solutions, pushing the boundaries of scientific understanding and engineering prowess.

Space Megastructures: Engineering Marvels Beyond Earth

In the grand tapestry of space colonization, the construction of colossal space megastructures emerges as a testament to human ingenuity and ambition. This section explores visionary concepts that stretch the limits of engineering and construction, envisioning vast habitats, ring worlds, and Dyson spheres that encircle stars. Architects of the future conceive habitats on a scale previously unimaginable, where entire cities float in the clouds of gas giants or orbit distant stars, harnessing their energy for human use.

Megastructures become hubs of innovation and human endeavor, fostering diverse ecosystems and serving as crucibles for scientific discovery. The construction techniques, materials, and societal challenges associated with these awe-inspiring creations are examined in detail, illuminating the path toward a future where space megastructures redefine the possibilities of human habitation and exploration.

Interstellar Colonization: Journey to the Stars

Beyond the confines of our solar system lie the distant shores of interstellar space, where exoplanets orbit alien stars, waiting to be explored and colonized. This section ventures into the realm of interstellar colonization, probing the theoretical concepts and propulsion technologies that could propel humanity to neighboring star systems. Scientists and theoreticians contemplate the challenges of relativistic travel, time dilation, and the vast cosmic distances that separate stars.

Concepts like the generation ships, fusion-powered starships, and laser propulsion systems become the focal points of interstellar exploration, where the dream of reaching new habitable worlds takes shape. Ethical considerations, the preservation of cultural identities during centuries-long journeys, and the impact of deep space travel on the human psyche are explored, painting a comprehensive picture of the profound journey toward the stars.

The Ethical Frontier: Guiding Principles in Space Colonization

In the boundless expanse of space, ethical considerations guide the path of human exploration and colonization. This section delves into the ethical dilemmas associated with space colonization, addressing issues such as the rights of potential extraterrestrial life forms, the preservation of planetary environments, and the responsibilities of human settlers in alien ecosystems. Ethicists, scientists, and policymakers collaborate to establish ethical frameworks that ensure the responsible stewardship of celestial bodies and the respectful coexistence with any indigenous life that may be encountered.

Discussions revolve around the equitable distribution of resources, the prevention of contamination in the search for extraterrestrial life, and the preservation of cultural and historical artifacts on alien worlds. Ethical guidelines become the foundation upon which humanity builds its legacy in the stars, shaping the collective conscience of interstellar pioneers.

Conclusion: The Tapestry of Cosmic Civilization

As this chapter concludes, the exploration of future horizons in space colonization paints a vivid tapestry of human potential and imagination. From the transformation of barren planets to the construction of colossal space megastructures and the epic odyssey toward interstellar destinations, humanity's journey beyond Earth transcends the boundaries of the known.

Ethical considerations, woven into the fabric of exploration, ensure that the legacy of space colonization is one of responsibility, respect, and understanding. The tapestry of cosmic civilization unfurls, revealing a future where the stars are not just distant pinpricks in the night sky but tangible destinations, where human endeavor, creativity, and the pursuit of knowledge create a legacy that spans the cosmos. As the chapter draws to a close, the horizon of possibilities expands, inviting future generations to continue the saga of space colonization, weaving new threads in the ever-expanding tapestry of cosmic exploration.

Chapter 11: The Environmental Challenges of Space Colonization

The Crucible of Sustainability: Agricultural Endeavors in Space

Within the vast, inhospitable realms of space, the cultivation of sustenance emerges as a paramount challenge for space colonies. This section navigates the intricate landscape of sustainable agriculture in extraterrestrial environments, where the absence of fertile soil and natural sunlight necessitates innovative solutions. Hydroponics, aeroponics, and advanced soilless farming techniques become the bedrock upon which colonies build their agricultural systems. Engineers and botanists collaborate to design closed-loop ecosystems, where waste products are transformed into nutrients, water is recycled, and artificial lighting mimics the sun's nurturing rays. The cultivation of crops, from nutrient-rich microgreens to staple food sources, becomes a testament to human adaptability and resilience, ensuring the nutritional well-being of colonists amidst the challenges of space.

Recycling the Cosmos: Waste Management and Resource Utilization

In the infinite expanse of space, the concept of waste takes on new dimensions, where every resource is precious and every material must be repurposed. This section delves into the intricacies of waste management in space colonies, exploring recycling technologies that extract value from discarded materials. From organic waste transformed into compost for agriculture to advanced recycling systems that reclaim water and metals, spacefaring societies master the art of resource conservation. The circular economy becomes a guiding principle, where waste is not a burden but a reservoir of potential resources. Engineers engineer closed-loop systems, where the byproducts of one process become the raw materials for another, minimizing the environmental footprint of space colonies and ensuring the sustainability of human presence in the cosmos.

Ecosystems in the Void: Biodiversity in Extraterrestrial Habitats

Biodiversity, the rich tapestry of life, finds itself redefined within the controlled confines of space colonies. This section explores the challenges and opportunities of fostering biodiversity in artificial environments, where microgravity and limited space influence the evolution of ecosystems. Biologists and ecologists study the behavior of organisms, from microorganisms vital to nutrient cycling to plants engineered for extraterrestrial life. Genetic diversity becomes a focus of research, ensuring the resilience of crops and animal species in the face of changing environmental conditions. Colonists, acting as stewards of these ecosystems, nurture fragile balances, striving to create harmonious habitats where life thrives. The study of extraterrestrial biodiversity offers insights into Earth's own ecosystems, illuminating the interconnectedness of life and the delicate balance required to sustain it in the harsh void of space.

Self-Contained Ecosystems: Nature in the Balance

Within the artificial confines of space colonies, the creation of self-contained ecosystems stands as a testament to humanity's mastery of nature. This section delves into the delicate balance required to maintain these enclosed biomes, where plants provide oxygen, recycle carbon dioxide, and purify water, while animals contribute to nutrient cycling and offer companionship to colonists. Engineers design habitats where nature and technology intertwine, fostering symbiotic relationships between organisms and the machines that sustain them.

From aquaponics systems where fish and plants coexist to the cultivation of edible algae in bioreactors, self-contained ecosystems become miniature reflections of Earth's biosphere. The challenge of balancing these ecosystems, ensuring the well-being of both colonists and the diverse life forms they nurture, becomes a hallmark of space colonization, embodying the harmonious coexistence of humanity and nature in the cosmic frontier.

Conclusion: The Green Frontier of Space

As this chapter concludes, the exploration of the environmental challenges of space colonization unveils a green frontier where innovation, adaptation, and ecological stewardship redefine humanity's relationship with the cosmos.

Through sustainable agriculture, efficient waste management, the cultivation of biodiversity, and the creation of self-contained ecosystems, spacefaring societies forge a path toward harmonious coexistence with the extraterrestrial environments they inhabit.

The lessons learned within the closed confines of space colonies echo back to Earth, inspiring sustainable practices and environmental consciousness on our home planet. The green frontier of space becomes a beacon of hope, showcasing humanity's ability to overcome challenges, embrace ingenuity, and nurture life amidst the cosmic void. As the chapter draws to a close, the vision of lush, thriving ecosystems in space becomes a reality, reminding us that even in the most barren reaches of the universe, life finds a way to flourish.

Chapter 12: Space Colonization and Climate Change

Harnessing the Cosmic Solution: Space-Based Solar Power

In the pursuit of sustainable energy solutions, space-based solar power emerges as a beacon of hope in the fight against climate change. This section explores the revolutionary concept of harvesting solar energy in space, where satellites equipped with vast solar arrays capture the unfiltered intensity of sunlight. The captured energy is then converted into microwaves or lasers and transmitted to receiving stations on Earth, providing a continuous and reliable source of clean energy.

Engineers and scientists collaborate to design intricate solar power satellites, utilizing advanced materials and efficient energy conversion technologies. The section delves into the advantages of space-based solar power, from its ability to operate continuously, unaffected by weather or nightfall, to its potential to meet the energy demands of entire nations. As space colonies pioneer this cosmic solution, Earth witnesses the dawn of a new era, where the boundless energy of the sun becomes humanity's ally in the battle against climate change.

Orchestrating Earth's Symphony: Carbon Capture Technologies

Amidst the rising concerns of carbon emissions, the orchestration of Earth's symphony takes center stage. This section explores innovative carbon capture technologies, where engineers and environmentalists collaborate to develop methods of capturing carbon dioxide from the atmosphere. From direct air capture systems that filter carbon dioxide from ambient air to bioenergy with carbon capture and storage (BECCS) initiatives that utilize plants to absorb carbon while generating energy, diverse approaches emerge.

Scientists engineer novel materials, such as metal-organic frameworks, capable of selectively capturing carbon dioxide molecules, paving the way for efficient carbon sequestration. The captured carbon dioxide finds purpose in enhanced oil recovery, the production of carbon-neutral fuels, and even the creation of valuable products, transforming a once harmful greenhouse gas into a valuable resource.

As Earth's atmosphere undergoes the delicate process of rebalancing, space colonies serve as testing grounds for these technologies, refining their efficacy and sustainability before they are deployed on a global scale.

Innovation Hubs in the Cosmic Outposts: Testing Grounds for Earth's Challenges

Space colonies, perched on the edge of the cosmic frontier, emerge not only as bastions of human ingenuity but also as innovation hubs addressing Earth's environmental woes. This section explores the role of space colonies as crucibles for developing and testing innovative solutions to Earth's environmental problems. Interdisciplinary teams of scientists, engineers, and visionaries collaborate within the controlled environments of colonies, experimenting with cutting-edge technologies.

From advanced waste recycling systems that mimic natural processes to sustainable agriculture practices that conserve water and promote biodiversity, space colonies become living laboratories. The section delves into the collaborative efforts between Earth and space, where knowledge flows seamlessly between these realms, accelerating the development of sustainable technologies.

As space colonies pioneer solutions for their own survival, Earth benefits from the fruits of this cosmic collaboration, ushering in a future where the challenges of climate change are met with unwavering resolve and boundless creativity.

Conclusion: The Cosmic Symbiosis

As this chapter concludes, the symbiotic relationship between space colonization and climate change mitigation comes into sharp focus. Space-based solar power illuminates Earth with clean, unending energy, while carbon capture technologies cleanse the atmosphere, heralding a new era of environmental responsibility. Simultaneously, space colonies serve as crucibles of innovation, birthing solutions to Earth's pressing challenges.

The cosmic symbiosis between humanity's ventures beyond Earth and the preservation of our home planet becomes a testament to our resilience, creativity, and determination. In the face of climate change, space colonization becomes more than a leap into the unknown; it becomes a lifeline, offering the wisdom and technologies necessary to safeguard Earth's future. As the chapter draws to a close, the vision of a harmonious coexistence between humanity, space, and Earth's delicate ecosystems becomes a reality, guiding us toward a future where the stars above and the planet below thrive in a cosmic dance of equilibrium.

Chapter 13: Space Medicine and Healthcare

The Pioneers of Space Medicine: Nurturing Health Beyond Earth

Within the cosmic confines of space colonies, a specialized realm of medicine emerges—space medicine. This section delves into the evolution of this pioneering field, tracing its roots from the early days of space exploration to the cutting-edge practices employed in modern space colonies. Space physicians, equipped with a profound understanding of human physiology and the challenges posed by microgravity, become the custodians of human health beyond Earth. The section explores the interdisciplinary nature of space medicine, where medical professionals collaborate with engineers, psychologists, and biologists to address the unique healthcare needs of spacefarers. From the meticulous selection of astronauts to the development of tailored exercise routines, space medicine pioneers a holistic approach to healthcare, ensuring the well-being of individuals embarking on cosmic journeys.

Telemedicine: Bridging the Gap Across Celestial Distances

Amidst the vast expanse of space, medical expertise transcends cosmic distances through the marvel of telemedicine. This section illuminates the role of telemedicine in space colonies, where advanced communication technologies connect astronauts with healthcare professionals on Earth. Through real-time video consultations, remote diagnostics, and teleoperated medical procedures, astronauts receive expert medical guidance regardless of the astronomical gulfs that separate them from Earth. Telemedicine becomes a lifeline, offering timely medical interventions and consultations, ensuring that health concerns, no matter how minor or complex, are addressed promptly. The section explores the challenges and triumphs of telemedicine, highlighting its vital role in preserving the health and well-being of space colonists, transforming the vast cosmic expanse into a realm where healthcare knows no bounds.

In the silent depths of space, where the void stretches infinitely, the heartbeat of medical care resonates through the marvel of telemedicine. Across the astronomical gulfs that separate space colonies from Earth, advanced communication technologies form the lifeline that connects astronauts with expert healthcare professionals. In this section, we dive deep into the transformative realm of telemedicine, exploring its pivotal role in preserving the health and well-being of space colonists, transcending the celestial distances that once seemed insurmountable.

The Digital Thread of Healing: Real-Time Video Consultations

Amidst the hum of life support systems and the distant twinkle of stars, astronauts find solace in the digital threads that bind them to Earth's medical expertise. Real-time video consultations, facilitated by high-bandwidth communication channels, allow astronauts to virtually meet healthcare professionals. Through these virtual encounters, physicians and specialists examine patients from millions of miles away, their expertise reaching across cosmic voids. Whether diagnosing a subtle anomaly or discussing the nuances of treatment plans, these consultations bring the healing touch of Earth to the confines of space, ensuring that medical concerns are met with precision and care.

Remote Diagnostics: Decoding Cosmic Ailments

In the microgravity of space, where every bodily signal takes on a new significance, the art of diagnostics becomes paramount. Remote diagnostic technologies, ranging from advanced medical imaging devices to biosensors integrated into spacesuits, provide a comprehensive view of astronauts' health. Analyzing data transmitted in real-time, healthcare professionals decode the subtlest of cosmic ailments. Whether it's monitoring cardiovascular health, assessing bone density, or scrutinizing physiological parameters affected by prolonged spaceflight, remote diagnostics offer insights that guide medical decisions, ensuring proactive interventions and personalized care tailored to the unique challenges of space.

Teleoperated Medical Procedures: Precision Across Light-Years

In the sterile environment of space clinics, where sterility meets the vast unknown, teleoperated medical procedures redefine the boundaries of surgical precision. Through robotic surgical systems controlled by skilled surgeons on Earth, intricate surgeries are performed with a level of precision that defies the vast cosmic distances. In this section, we delve into the intricacies of teleoperated surgeries, exploring how advanced robotics, haptic feedback systems, and augmented reality interfaces enable surgeons to operate on astronauts with unparalleled accuracy. From minor procedures to complex surgeries, teleoperated medical interventions ensure that space colonists receive the best possible care, even in the face of surgical challenges unique to extraterrestrial environments.

Challenges and Triumphs: Navigating the Telemedical Cosmos

While telemedicine bridges the gap between space and Earth, it is not without its challenges.

Communication delays, limited bandwidth, and the complexities of performing procedures in microgravity pose formidable obstacles. Yet, in overcoming these challenges, telemedicine stands as a triumph of human ingenuity. This section explores the hurdles faced in the telemedical cosmos, shedding light on the innovations that overcome these challenges. From adaptive communication protocols to cutting-edge robotic interfaces, each triumph represents a step toward ensuring that the vastness of space does not compromise the quality of healthcare delivered to space colonists.

In the cosmic expanse, telemedicine becomes the embodiment of compassion and expertise, transcending the barriers of time and space. Through real-time consultations, remote diagnostics, and teleoperated surgeries, the art of healing reaches across light-years, ensuring that the health of space colonists remains safeguarded. In the quiet corridors of space clinics, the digital pulse of telemedicine echoes, reminding astronauts that, no matter how far they roam, the expertise of Earth's healers is always within reach, a testament to the boundless potential of human innovation in the face of cosmic challenges.

The Challenge of Healing in Microgravity: Surgical Precision in Weightlessness

In the microgravity ballet of space colonies, performing medical procedures becomes a delicate art. This section delves into the challenges posed by microgravity, where traditional surgical techniques are reimagined and refined to accommodate the weightless environment. Surgeons, equipped with state-of-the-art tools and augmented reality systems, navigate the complexities of medical procedures with precision and finesse. From minor interventions to complex surgeries, space physicians pioneer innovative techniques, ensuring the safety and efficacy of medical procedures in cosmic settings. The section explores the development of specialized medical equipment, such as microsurgical robots and advanced imaging systems, designed to enhance the capabilities of space surgeons. With unwavering determination and ingenuity, medical professionals in space colonies redefine the boundaries of healthcare, demonstrating that the absence of gravity does not impede the pursuit of medical excellence.

In the graceful ballet of microgravity, where every movement is choreographed by the absence of Earth's pull, the art of healing takes on a new dimension. This section delves into the intricacies of performing medical procedures in the ethereal realm of space colonies. Here, the challenges posed by microgravity are met with unwavering determination and surgical precision, redefining traditional techniques to accommodate the weightless environment.

The Dance of Precision: Reimagining Surgical Techniques

In the sterile confines of space clinics, surgeons become choreographers, orchestrating delicate procedures in a dance devoid of gravitational constraints. Traditional surgical techniques are reimagined, refined, and adapted to the unique challenges of microgravity. Every movement, every incision, is executed with utmost precision, guided by augmented reality systems that overlay critical information onto surgeons' field of vision. In this weightless ballet, surgeons perform with finesse, ensuring the safety and efficacy of medical procedures. From suturing minor wounds to conducting intricate surgeries, space physicians master the dance of precision, showcasing the fusion of expertise and innovation that defines the art of healing in microgravity.

Tools of the Trade: State-of-the-Art Equipment for Cosmic Surgeries

Equipped with cutting-edge tools, space surgeons navigate the complexities of medical interventions with grace. Microsurgical robots, guided by skilled hands, enable procedures of unparalleled precision. These robotic assistants, designed specifically for the intricate demands of space surgeries, work in harmony with surgeons, enhancing their capabilities and ensuring the success of delicate operations. Advanced imaging systems, utilizing technologies like magnetic resonance imaging (MRI) and three-dimensional ultrasound, provide surgeons with detailed insights into patients' anatomy. With these tools at their disposal, medical professionals in space colonies push the boundaries of healthcare, exploring new frontiers in surgical excellence.

Innovations Beyond Earth: Redefining Medical Boundaries

Amidst the silent hum of life support systems, medical professionals in space colonies pioneer innovations that reverberate across cosmic distances. In this section, we explore the development of specialized medical equipment designed exclusively for extraterrestrial healthcare. From self-sterilizing surgical instruments to biofeedback systems that monitor patients' vital signs in real-time, these innovations redefine the standards of medical care. Space physicians, driven by the spirit of exploration and the pursuit of excellence, continue to push the envelope of medical science, demonstrating that the absence of gravity does not impede the quest for medical perfection.

The Legacy of Healing: Inspiring Future Generations

As surgeons perform their weightless ballet, they leave behind a legacy of healing that echoes through the corridors of space colonies. Their innovations, their dedication, inspire future generations of medical professionals, both on Earth and among the stars. This section illuminates the enduring impact of space surgeries, highlighting how the challenges of microgravity catalyze advancements that ultimately benefit healthcare practices on Earth. From refined techniques to state-of-the-art equipment, the legacy of healing in space becomes a beacon, guiding the evolution of medical science and inspiring a new era of surgical excellence.

In the cosmic theater of microgravity, where the absence of weight challenges the very essence of healing, surgeons redefine the boundaries of their art. With surgical precision, unwavering determination, and the aid of cutting-edge technology, they navigate the challenges of microgravity, demonstrating that the pursuit of medical excellence knows no bounds. As they choreograph the dance of healing in space, they inspire generations, leaving an indelible mark on the history of medicine, both terrestrial and cosmic.

The Mind-Body Connection: Nurturing Mental Health in Isolation

Isolation and confinement, inherent to space exploration, cast profound shadows on mental health. This section delves into the psychological aspects of healthcare in space colonies, exploring the impact of prolonged isolation on astronauts' mental well-being. Psychologists and counselors, armed with insights from terrestrial behavioral sciences, pioneer therapeutic interventions tailored for the unique challenges of space. Virtual reality therapy, mindfulness practices, and personalized counseling sessions become essential tools in nurturing mental resilience. The section illuminates the strategies employed to foster a supportive environment, where camaraderie, open communication, and recreational activities serve as anchors for mental well-being. As space physicians address the intricate interplay between the mind and the cosmos, space colonies become sanctuaries of mental health, ensuring that astronauts embark on their celestial odysseys with not only physical but also psychological strength.

In the vast expanse of space, where silence reigns and stars stretch infinitely, the human mind faces its most profound challenge: isolation. In the heart of space colonies, where astronauts dwell in confined quarters, the psychological well-being of individuals becomes paramount. This section delves deep into the intricate interplay between the mind and the cosmos, exploring the innovative approaches employed to nurture mental health amidst the solitude of space.

The Shadows of Isolation: Understanding the Psychological Strain

Isolation and confinement cast long shadows, touching the very core of astronauts' mental resilience. Prolonged space missions demand an acute understanding of the psychological strain experienced by those who venture beyond Earth's embrace. Psychologists and counselors, armed with insights from terrestrial behavioral sciences, delve into the complexities of human emotions in the isolated realms of space. From the longing for familiar landscapes to the subtle echoes of homesickness, these professionals become the silent guardians of astronauts' mental well-being, offering tailored support to navigate the labyrinthine corridors of the mind.

Therapeutic Interventions: Tools for Mental Resilience

In the arsenal of mental health professionals, a diverse array of therapeutic interventions emerges, each designed to foster mental resilience in the face of isolation. Virtual reality therapy becomes a portal, transporting astronauts back to Earth, allowing them to momentarily escape the cosmic solitude. Mindfulness practices, rooted in ancient wisdom, find their place amidst the stars, offering moments of tranquility amid the cosmic chaos. Personalized counseling sessions, conducted with empathy and understanding, become lifelines, providing astronauts with spaces where they can share their thoughts, fears, and aspirations. These interventions, tailored for the unique challenges of space, become beacons of hope, illuminating the psychological landscape of astronauts and ensuring that they traverse the cosmic void with strength and stability.

Building Supportive Environments: Camaraderie, Communication, and Recreation

Within the confined walls of space colonies, supportive environments become sanctuaries for astronauts' mental health. Camaraderie flourishes, creating bonds that transcend the vastness of space. Open communication becomes a cornerstone, allowing astronauts to express their thoughts and emotions freely, fostering a sense of connection amidst the cosmic solitude. Recreation, in various forms, becomes an essential component, providing avenues for relaxation and creative expression. From musical jam sessions that echo through the corridors to art installations that capture the cosmic wonder, recreational activities serve as anchors, grounding astronauts in the realm of the familiar.

Psychological Strength: The Foundation of Celestial Odysseys

As astronauts embark on their celestial odysseys, they carry not only the physical weight of their missions but also the psychological strength nurtured within the supportive embrace of space colonies. The mind-body connection, once perceived as a challenge, transforms into a source of resilience. In the face of isolation, astronauts discover the depths of their mental fortitude, emerging as pioneers of the mind, exploring uncharted territories within themselves. In the cosmic tapestry of space exploration, the nurturing of mental health becomes an essential thread, weaving through the very fabric of human endurance and determination, ensuring that the human spirit continues to soar amidst the stars.

Conclusion: Health Beyond Horizons

As this chapter concludes, the tapestry of space medicine and healthcare unfolds, depicting a future where human health transcends terrestrial boundaries. From the meticulous care provided by space physicians to the seamless connections forged through telemedicine, space colonies become bastions of well-being, nurturing the mind and body of every cosmic voyager. In the face of isolation and the challenges of microgravity, healthcare in space colonies emerges as a beacon of hope, showcasing humanity's resilience and adaptability. The chapter concludes with a vision of health beyond horizons, where the lessons learned in the cosmic crucible not only benefit spacefarers but also inspire transformative advancements in healthcare on Earth. As the stars above watch over the pioneers of space medicine, a new era dawns— one where health knows no cosmic confines, and the human spirit soars boundlessly into the cosmic unknown.

Chapter 14: Education and Research in Space

The Cosmic Classroom: Space Colonies as Centers of Learning

Within the domes and modules of space colonies, a cosmic revolution in education unfolds. This section delves into the establishment of space universities and research institutions, where the pursuit of knowledge transcends terrestrial boundaries. Educators and researchers, drawn by the allure of the cosmos, converge in these celestial hubs, transforming space colonies into vibrant centers of learning. The section explores the innovative pedagogical approaches employed in space-based education, where virtual reality simulations, interactive experiments, and immersive learning experiences captivate the minds of students. From astronomy lectures under the gleaming stars to engineering workshops amidst advanced spacecraft, space universities redefine the concept of the classroom, inspiring the next generation of scientists, engineers, and explorers.

In the cosmic tapestry of space colonization, education emerges as a guiding star, illuminating the path toward a future where knowledge knows no earthly bounds. Within the domes and modules of space colonies, a cosmic revolution in education unfolds, reshaping the very essence of learning. This section delves into the establishment of space universities and research institutions, where the pursuit of knowledge becomes a celestial endeavor.

Educational Odyssey: Beyond Terrestrial Boundaries

In the hallowed halls of space universities, a diverse community of educators and researchers converges, drawn by the magnetic allure of the cosmos. Here, the boundaries of terrestrial education dissolve, paving the way for an educational odyssey that spans the universe. Astronomy lectures, conducted under the gleaming stars of distant galaxies, become windows into the cosmic wonders that surround space colonies. Engineering workshops, set amidst the intricate design of advanced spacecraft, transform theoretical knowledge into tangible innovation.

Innovative Pedagogy: The Marriage of Technology and Learning

In the vast expanse of space, education embraces innovation with open arms. Virtual reality simulations, powered by cutting-edge technology, become portals to uncharted worlds, allowing students to explore alien landscapes and unravel the mysteries of the universe. Interactive experiments, conducted in state-of-the-art laboratories, come to life, engaging students in scientific inquiry that transcends the confines of textbooks. Immersive learning experiences, where the boundaries between reality and imagination blur, captivate the minds of students, inspiring them to reach for the stars.

Inspiring the Next Generation: Nurturing Future Scientists and Explorers

Within the cosmic classrooms of space colonies, a new generation of scientists, engineers, and explorers takes its first steps. The inspirational stories of space pioneers echo through the corridors, igniting the spark of curiosity in young minds. Encounters with seasoned researchers become transformative moments, shaping the aspirations of future astronomers, biologists, and physicists. The legacy of space exploration becomes a beacon, guiding the educational journey of students, urging them to dream beyond the sky and reach for the infinite possibilities that the cosmos holds.

Beyond Earthly Horizons: The Cosmic Legacy of Space-Based Education

As the pages of space-based education unfold, they create a cosmic legacy that transcends earthly horizons. The knowledge cultivated within the celestial halls of space universities becomes a precious gift, enriching humanity's understanding of the universe. In the ever-expanding cosmos, the pursuit of knowledge becomes a timeless odyssey, with space colonies serving as beacons of enlightenment, illuminating the cosmic darkness with the brilliance of human intellect. As students graduate from the cosmic classrooms, they carry with them not only degrees but also the boundless potential to explore, discover, and redefine the boundaries of human knowledge, ensuring that the legacy of space-based education continues to echo through the corridors of time.

Beyond the Horizon: Unraveling Cosmic Mysteries through Research

In the laboratories of space colonies, scientists embark on unprecedented voyages of discovery. This section illuminates the unique scientific opportunities afforded by the space environment, where researchers delve into fields such as astronomy, materials science, and fundamental physics. The absence of atmospheric distortion and the interference of city lights unveil the cosmic canvas in its purest form, enabling astronomers to unravel the mysteries of distant galaxies and explore the fabric of the universe. Materials scientists, liberated from gravitational constraints, pioneer advancements in materials with extraordinary properties, poised to revolutionize industries on Earth. The section delves into groundbreaking experiments conducted in space, from studies on the behavior of fluids in microgravity to investigations into the fundamental forces shaping the cosmos. Space colonies become crucibles of scientific innovation, where the pursuit of knowledge knows no celestial bounds.

In the laboratories of space colonies, where the hum of scientific inquiry resonates through the sterile air, researchers embark on unprecedented voyages of discovery. Here, the boundaries of scientific exploration extend beyond the terrestrial realm, opening portals to the vast cosmic expanse. This section illuminates the unique scientific opportunities afforded by the space environment, where the pursuit of knowledge takes on a celestial significance.

Pure Cosmic Canvas: Astronomy Unveiled

In the absence of atmospheric distortion and the interference of city lights, astronomers within space colonies gaze upon the night sky in its unadulterated brilliance. The cosmic canvas unfolds in its purest form, revealing the secrets of distant galaxies and the enigmatic dance of celestial bodies. Observatories equipped with state-of-the-art telescopes become gateways to the universe, allowing scientists to unravel cosmic mysteries that have eluded terrestrial observations. Here, the very fabric of the universe is explored, as astronomers chart the evolution of galaxies, study the birth and death of stars, and peer into the depths of black holes.

Materials Beyond Earth: Pioneering Innovations

In the weightless embrace of space, materials scientists redefine the limits of innovation. Liberated from the constraints of gravity, they pioneer advancements in materials with extraordinary properties, poised to revolutionize industries on Earth. Nanostructures, meticulously engineered, exhibit unparalleled strength and conductivity, promising groundbreaking developments in electronics and energy technologies. Composites, crafted under the unique conditions of microgravity, boast exceptional durability, offering solutions to challenges in transportation and infrastructure. The laboratories within space colonies become crucibles of material science, where the boundaries of what is possible are constantly pushed, leading to innovations that have far-reaching implications for humanity's technological future.

Fluid Dynamics in Microgravity: Unraveling Fundamental Forces

In the microgravity ballet of space colonies, the behavior of fluids takes on a mesmerizing complexity. Scientists delve into the intricacies of fluid dynamics, studying phenomena that are impossible to replicate under Earth's gravity. Droplets of water form perfect spheres, clinging to surfaces in defiance of gravity's pull. Buoyancy-driven convection gives way to peculiar patterns, illuminating the fundamental forces shaping the behavior of fluids in the absence of gravity. Experiments within space laboratories probe the mysteries of fluid behavior, shedding light on phenomena that have practical applications on Earth, from improved fuel efficiency to enhanced drug delivery systems.

Cosmic Crucibles of Innovation: The Pursuit of Boundless Knowledge

Within the laboratories of space colonies, the pursuit of knowledge knows no celestial bounds. Scientists, fueled by curiosity and guided by precision, conduct groundbreaking experiments that unravel the mysteries of the cosmos. From the study of exotic materials to the exploration of fundamental forces, space colonies become cosmic crucibles of innovation. Here, the human quest for understanding transcends earthly confines, reaching toward the infinite possibilities that the universe holds. In these laboratories, the future of technology and scientific discovery is forged, ensuring that the legacy of space-based research becomes a beacon, illuminating the path toward a future where the boundaries of knowledge are pushed ever further, enriching humanity's understanding of the universe and the wonders it contains.

Challenges and Innovations: Space-Based Education in the Cosmic Frontier

The vastness of space presents both challenges and innovations in education. This section addresses the intricacies of space-based education, exploring the hurdles faced by educators and students in the cosmic frontier. From the adaptation of curricula to accommodate the unique space environment to the development of specialized educational tools, educators pioneer transformative solutions. Virtual classrooms bridge the gaps between Earth and space, connecting students across continents with astronauts and scientists conducting research in real time. The evolution of space-based educational technologies, from augmented reality experiments to interactive online platforms, fostering a global community of learners united by their passion for the cosmos. Through collaborative initiatives and international partnerships, space-based education becomes a beacon of unity, transcending terrestrial boundaries and nurturing a generation of curious minds eager to explore the wonders of the universe.

In the boundless expanse of space, education undergoes a transformative journey, presenting both challenges and innovative solutions. This section delves into the intricacies of space-based education, exploring the hurdles faced by educators and students in the cosmic frontier. Amidst the stars and celestial wonders, a new paradigm of learning emerges, shaping the minds of future pioneers.

Adaptation and Transformation: Curricula in the Cosmic Classroom

Adapting curricula to the unique demands of the space environment becomes a paramount challenge for educators. Traditional subjects find new dimensions as science, mathematics, and humanities intertwine with the mysteries of the cosmos. In this section, we explore the innovative ways in which educators transform conventional curricula, infusing them with the wonders of space exploration. From integrating astrophysics into mathematics lessons to exploring the philosophical implications of space travel, the cosmic classroom becomes a crucible of interdisciplinary learning. Here, students embark on intellectual journeys that transcend the confines of individual subjects, fostering holistic understandings of the universe and humanity's place within it.

Educational Tools: The Evolution of Space-Based Learning

At the forefront of space-based education are the tools that bridge the gap between Earth and the cosmic frontier. This section delves into the evolution of educational technologies, from the pioneering days of space exploration to the cutting-edge innovations of the present. Augmented reality experiments transport students to the surfaces of distant planets, allowing them to conduct virtual geological explorations. Interactive online platforms create global classrooms, where students collaborate on experiments and projects, fostering a sense of camaraderie that transcends national borders. From immersive virtual reality simulations of astronomical phenomena to collaborative coding projects that design space probes, educational tools become gateways to the wonders of the universe, inspiring the next generation of scientists, engineers, and explorers.

Connecting Continents: Global Collaboration in Space-Based Education

In the interconnected world of space-based education, international collaboration becomes a cornerstone of learning. This section illuminates the collaborative initiatives that unite students and educators across continents, fostering a global community of learners bound by their shared passion for the cosmos. Virtual classrooms dissolve geographical distances, allowing students from diverse cultures to engage in real-time discussions with astronauts and scientists conducting research in space. Joint experiments, conducted simultaneously in multiple countries, inspire a spirit of friendly competition and mutual discovery. Through international partnerships, students exchange perspectives, cultural nuances, and scientific insights, enriching their educational experiences and nurturing a sense of global citizenship.

Unity in Exploration: Nurturing Curiosity Across Borders

In the cosmic classroom, unity blossoms from curiosity. This section celebrates the shared enthusiasm for exploration that transcends terrestrial boundaries. Students, regardless of their geographical locations, become cosmic explorers, embarking on quests for knowledge that traverse the vastness of space. Collaborative projects unravel the mysteries of the universe, encouraging innovative thinking and problem-solving skills. International competitions inspire friendly rivalries, igniting the spirit of discovery in young minds. Through space-based education, a generation of curious thinkers emerges, unified by their shared wonder for the cosmos. As they gaze at the stars, they see not only distant suns and galaxies but also the infinite possibilities of human potential. In this unity of exploration, the future of humanity's cosmic endeavors finds its foundation, nurtured by the collective curiosity of an interconnected world.

Conclusion: Knowledge Without Borders

As this chapter concludes, the narrative of education and research in space unfolds as a testament to humanity's insatiable curiosity and thirst for knowledge. Space colonies, once mere habitats, evolve into bastions of learning, where the brightest minds converge to unravel cosmic mysteries and inspire future generations.

The challenges overcome and the innovations pioneered in the cosmic frontier echo a resounding message: knowledge knows no borders. In the celestial classrooms of space colonies, a new era of enlightenment dawns, where the collective wisdom of humanity illuminates the cosmic expanse. As students gaze at distant stars and researchers probe the depths of the universe, they do so not as isolated individuals but as integral parts of a global community united by the pursuit of understanding. The chapter concludes with a vision of knowledge without borders, where the lessons learned in the cosmic crucible enrich the tapestry of human understanding, paving the way for a future where the cosmos becomes an infinite reservoir of wisdom and discovery.

Chapter 15: Space Tourism and Recreation

The Celestial Playground: Space Tourism and Its Promise

In the heart of the cosmos, a new frontier beckons thrill-seekers and adventurers. This section explores the burgeoning realm of space tourism, where Earthbound tourists transcend the confines of our planet to experience the awe-inspiring wonders of space. The chapter begins by delving into the evolution of space tourism, from its nascent stages to a burgeoning industry that promises celestial vistas to those with wanderlust. It illuminates the various forms of space tourism, from suborbital flights offering fleeting glimpses of Earth from the edge of space to orbital journeys that immerse travelers in the mesmerizing dance of planets and stars.

The section unravels the allure of space resorts, where luxurious accommodations meld seamlessly with cosmic vistas, offering guests a chance to float weightlessly amidst the stars. Through vivid descriptions and immersive narratives, readers are transported to a future where space tourism becomes an accessible dream, inviting individuals from all walks of life to embark on transformative journeys beyond Earth's atmosphere.

The Evolution of Space Tourism: From Dream to Reality

In the boundless expanse of the cosmos, a revolutionary venture unfolds: space tourism, where the adventurous spirit of humanity meets the wonders of the universe. This section delves into the evolution of space tourism, tracing its inception from the realm of science fiction to the tangible reality of the present day. As Earthbound explorers cast their eyes to the stars, they yearn for more than just terrestrial landscapes. The allure of space becomes a magnetic force, pulling them toward the ultimate frontier of human experience.

Suborbital Adventures: Brief Encounters with Cosmic Majesty

At the threshold of Earth's atmosphere, suborbital spaceflights offer intrepid travelers a fleeting taste of the cosmos. This section illuminates the suborbital experience, where spacecraft breach the boundaries of our planet's atmosphere, granting passengers a breathtaking panorama of the Earth from the edge of space. With weightlessness embracing them and the curvature of the Earth painting a mesmerizing picture below, tourists become astronauts for a brief, exhilarating moment. Through firsthand accounts and vivid descriptions, readers are transported aboard these pioneering vessels, where the blue hues of Earth blend seamlessly with the cosmic void.

Orbital Journeys: Dancing Among the Stars

For those with an insatiable thirst for celestial grandeur, orbital spaceflights offer immersive journeys amidst the cosmic ballet of planets and stars. This part of our exploration delves into the intricate choreography of orbital missions, where spacecraft glide gracefully around the Earth, revealing the majesty of our planet against the backdrop of the infinite cosmos. Astronauts-turned-guides share the wonders of weightlessness, inviting tourists to float effortlessly, experiencing the cosmic perspective that has inspired generations of stargazers. The section unveils the sensory symphony of orbital travel, where the brilliance of stars, the serenity of space, and the fragility of Earth converge to create an unparalleled cosmic ballet.

Luxury in the Stars: The Allure of Space Resorts

In the celestial playground, space resorts emerge as sanctuaries of luxury and tranquility. This segment delves into the architectural marvels that define space resorts, where opulent accommodations merge seamlessly with the cosmic panorama. Guests find themselves in floating havens, surrounded by panoramic views of distant galaxies and nebulae. Here, weightlessness becomes a norm, and Earthly constraints fade into insignificance. With attentive staff catering to every need, visitors experience unparalleled comfort, indulging in gourmet space cuisine and savoring the awe-inspiring beauty of the universe from the comfort of their celestial abodes. Through immersive narratives and detailed descriptions, readers step into these cosmic retreats, where the promise of transformative experiences awaits.

An Accessible Dream: Inviting All to the Celestial Soiree

As space tourism advances, it becomes a dream accessible to people from all walks of life. This section explores the democratization of space travel, where the boundaries of financial constraints blur, allowing diverse individuals to embark on transformative journeys beyond Earth. Innovations in technology and business models pave the way for a future where space tourism becomes a reality for enthusiasts, scientists, artists, and families alike. The section highlights initiatives that bridge the gap between aspiration and experience, making the cosmic soiree an inclusive celebration. With each passing day, the celestial playground becomes more inviting, beckoning humanity to transcend the confines of our world and step into the vast expanse of the unknown.

Beyond Gravity: Leisure Activities in Microgravity

In the microgravity environment of space, leisure takes on a mesmerizing dimension. This section dives into the myriad recreational activities that await space tourists, from exhilarating spacewalks that allow guests to navigate the vastness of space to serene moments of stargazing within transparent domes. Readers are introduced to the innovative leisure facilities of space resorts, where zero-gravity sports, holographic gaming, and immersive art installations redefine the concept of entertainment. Aboard space stations orbiting high above Earth, tourists partake in activities that bridge the gap between terrestrial pastimes and cosmic experiences. From culinary delights prepared with cosmic ingredients to concerts that resonate through the metal walls of spacecraft, every moment becomes a celebration of human creativity and ingenuity. The section paints a vivid picture of life in microgravity, where the absence of gravity transforms ordinary activities into extraordinary adventures, inviting tourists to revel in the boundless possibilities of the cosmic playground.

Boundless Leisure in Microgravity: The Cosmic Playground Unveiled

In the mesmerizing realm of microgravity, leisure transcends the limits of Earthly constraints, inviting space tourists to embark on a celestial odyssey unlike any other. This section immerses readers in the kaleidoscope of recreational activities that define the cosmic playground, where every moment becomes a celebration of human imagination and adaptability.

Spacewalks: Navigating the Cosmic Vistas

One of the most exhilarating experiences in space tourism is the spacewalk, where tourists don specially designed suits and venture into the cosmic abyss. This part of our exploration delves into the intricacies of spacewalks, where tourists become cosmic explorers, floating amidst the stars, unbound by the pull of gravity. With Earth suspended below and the infinite expanse above, spacewalks offer a profound sense of freedom and awe. Astronaut guides lead tourists through the silent void, revealing the secrets of the cosmos while ensuring safety in the vacuum of space. Through immersive descriptions, readers accompany space tourists on these transformative journeys, where the vastness of space becomes a canvas for the most extraordinary adventure of a lifetime.

Stargazing in Transparent Domes: Celestial Serenity

Within the transparent domes of space resorts, tourists are treated to the serenity of stargazing. This segment illuminates the experience of stargazing in space, where the twinkling stars and distant galaxies paint a mesmerizing tableau against the cosmic canvas. In these tranquil environments, tourists find solace in the beauty of the universe, their gazes unobstructed by atmospheric disturbances. Astronomers guide their observations, pointing out constellations, planets, and celestial phenomena, turning stargazing into a cosmic journey of discovery. The section captures the essence of celestial serenity, where tourists connect with the cosmos in the most intimate and profound manner, their thoughts drifting among the stars.

Zero-Gravity Sports and Holographic Gaming: Thrills Beyond Earth

Space resorts redefine the concept of sports and gaming, offering tourists an array of activities that defy the boundaries of gravity. This portion of our exploration delves into zero-gravity sports, where tourists engage in activities like floating basketball or weightless swimming, experiencing the thrill of athleticism in a realm devoid of gravitational pull. Simultaneously, holographic gaming brings virtual worlds to life, allowing tourists to immerse themselves in fantastical adventures within the confined spaces of spacecraft. With augmented reality headsets, players traverse virtual landscapes, solving puzzles, engaging in epic battles, and experiencing narratives that blur the lines between reality and imagination. Through detailed accounts, readers step into these virtual arenas, where the boundaries between the physical and the digital dissolve, offering an unparalleled blend of excitement and innovation.

Art Installations and Culinary Delights: The Fusion of Creativity and Gastronomy

In the microgravity environment, art and cuisine become extraordinary expressions of human creativity. This part of our exploration delves into the immersive art installations adorning space resorts, where artists craft mesmerizing visual experiences that come to life in weightlessness. From floating sculptures to interactive light displays, the cosmic environment becomes a canvas for artistic innovation, captivating the senses and igniting the imagination. Simultaneously, culinary delights take on a celestial flair, as chefs prepare dishes with cosmic ingredients, transforming meals into multisensory experiences. Guests indulge in gourmet delicacies, savoring the flavors of Earth and space, their palates tantalized by the fusion of creativity and gastronomy. Through evocative descriptions, readers are invited to taste the celestial cuisine and immerse themselves in the artistic wonders of space, where every meal and art installation becomes a testament to the boundless ingenuity of humanity.

Concerts in the Cosmos: Harmonies Beyond the Stars

The harmony of music finds new resonance in the microgravity environment, where concerts in space become ethereal experiences. This segment explores the magic of space concerts, where musicians and performers from diverse cultures and genres come together to create transcendent musical performances. Aboard space stations, sound waves travel differently, filling the metallic chambers with melodies that resonate in the hearts of listeners. Musicians strum guitars, play keyboards, and sing, their music echoing through the halls of spacecraft, transforming the sterile environment into a symphony of human expression. With the absence of gravity, dancers and performers move gracefully, unhindered by Earthly constraints, adding a visual spectacle to the auditory delight. Through detailed narratives, readers attend these cosmic concerts, feeling the vibrations of music that transcends terrestrial boundaries, inviting everyone to join in the celestial celebration of human creativity.

The Cosmic Playground: Where Imagination Soars Without Limits

In the heart of microgravity, the cosmic playground becomes a sanctuary of boundless possibilities, where every activity, every moment, and every experience transcends the ordinary. This section paints a vivid picture of life in microgravity, where space tourists revel in the thrill of spacewalks, find solace in stargazing, engage in zero-gravity sports and holographic adventures, savor exquisite culinary creations, immerse themselves in captivating art installations, and lose themselves in the harmonies of space concerts. Each endeavor becomes a celebration of human ingenuity, a testament to the capacity of the human spirit to adapt, innovate, and create amidst the challenges of the cosmic frontier. As tourists float weightlessly, their dreams take flight, soaring among the stars, reflecting the boundless imagination of humanity in the vast expanse of the universe.

Ethical Considerations: Balancing Exploration and Preservation

As space tourism expands its horizons, ethical considerations come to the forefront. This section navigates the complex ethical landscape of space tourism, addressing questions of environmental impact, resource allocation, and cultural preservation. It delves into the responsible practices adopted by space tourism operators to minimize their ecological footprint, ensuring the preservation of celestial environments for future generations. The chapter explores the delicate balance between exploration and conservation, emphasizing the need for sustainable approaches that safeguard both the wonders of space and the cultural heritage of celestial bodies. It contemplates the ethical dimensions of lunar and Martian tourism, where ancient landscapes and potentially habitable environments pose unique challenges to ethical tourism practices. Through thought-provoking discussions and ethical dilemmas, readers are encouraged to contemplate the future of space tourism, envisioning a future where the wonders of the cosmos are accessible to all while preserving the sanctity of unexplored worlds.

Balancing Ethical Frontiers: Navigating the Cosmos Responsibly

In the expanding realm of space tourism, ethical considerations stand as guiding beacons, illuminating the path toward responsible cosmic exploration. This section embarks on a profound journey through the ethical landscape of space tourism, exploring the intricate web of dilemmas and challenges that arise when humanity ventures into the celestial expanse. With a careful balance between curiosity and conservation, space tourism operators grapple with ethical questions that redefine our understanding of exploration, conservation, and cultural preservation in the cosmic frontier.

Preservation in the Stars: Guardians of Celestial Heritage

One of the paramount ethical considerations in space tourism is the preservation of celestial environments. This part of our exploration delves into the responsible practices adopted by space tourism operators to safeguard the sanctity of unexplored worlds. Ethical guidelines and protocols are meticulously crafted, ensuring that lunar, Martian, and other celestial landscapes remain untouched by human interference. Space tourists are educated on the importance of minimal impact, as they traverse ancient terrains and gaze upon extraterrestrial wonders. The section explores innovative technologies employed to minimize ecological footprints, from advanced waste recycling systems to eco-friendly habitats that blend seamlessly with the natural cosmic surroundings. Through detailed examinations, readers gain insight into the meticulous efforts made to preserve the celestial heritage, safeguarding the mysteries of space for generations yet unborn.

Ethical Tourism Practices: Striking a Balance Between Wonder and Respect

Space tourism confronts unique ethical dilemmas, especially concerning encounters with ancient landscapes and potentially habitable environments. This segment delves into the

119

nuanced ethical considerations of lunar and Martian tourism, where the allure of exploring potentially life-supporting environments must be delicately balanced with respect for extraterrestrial habitats. Ethical tourism practices are explored, outlining the dos and don'ts for space tourists as they tread on alien terrain. From preserving ancient geological formations to respecting potential signs of past life, ethical guidelines serve as moral compasses, ensuring that the wonders of the cosmos are marveled at with awe and reverence. Through thought-provoking scenarios and ethical debates, readers are challenged to contemplate their roles as cosmic tourists, inspiring a profound sense of responsibility and respect for the celestial realms they encounter.

The Future of Ethical Space Tourism: A Vision of Harmony

As space tourism advances, a vision of ethical harmony emerges on the cosmic horizon. This section paints a picture of a future where ethical considerations are woven into the very fabric of space exploration. Responsible tourism practices, cultural preservation efforts, and environmental conservation become integral components of cosmic adventures. The chapter explores visionary initiatives, from international agreements that establish protected cosmic zones to educational campaigns that promote ethical awareness among space tourists. Through futuristic scenarios and ethical foresight, readers glimpse a future where humanity explores the cosmos with a deep sense of wonder and a profound respect for the celestial wonders they encounter. The section inspires contemplation, encouraging readers to envision a future where ethical considerations serve as guiding stars, leading humanity toward a harmonious coexistence with the wonders of the universe.

Conclusion: The Odyssey Continues

As this chapter concludes, readers are left with a sense of wonder and anticipation for the future of space tourism. The celestial playground, once reserved for astronauts and scientists, opens its gates to humanity, inviting individuals to embark on journeys that redefine the boundaries of exploration and leisure. The chapter's narrative weaves together the thrill of space travel, the enchantment of cosmic experiences, and the ethical considerations that accompany humanity's foray into the stars. It paints a vision of a future where space tourism becomes a catalyst for scientific discovery, cultural exchange, and the shared dreams of a global community. The odyssey continues, with space tourism serving as a gateway to the cosmos, where the mysteries of the universe become not just the realm of scientists but the shared heritage of all humanity. As tourists gaze at Earth from orbit and immerse themselves in the marvels of distant worlds, they do so not merely as spectators but as active participants in the epic tale of human exploration, reminding the world that the final frontier is, and always will be, within reach.

Chapter 16: Space Colonies and International Relations

The New Frontier: Geopolitical Implications of Space Colonization

In the boundless expanse of space, the seeds of geopolitical dynamics are sown anew. This section delves deep into the intricate web of international relations that emerge as humanity ventures beyond Earth's confines. It explores the geopolitical implications of space colonization, unraveling the complex interplay of collaboration, competition, and diplomacy among nations. From the collaborative efforts of space agencies to the strategic rivalries that echo the geopolitical landscape of Earth, readers are immersed in a world where the stakes are as high as the cosmic heavens. The chapter traces the evolution of international space partnerships, from joint missions that bridge cultural divides to the establishment of multinational space stations that serve as symbols of global cooperation. It sheds light on the delicate balance between shared scientific goals and national interests, where the pursuit of knowledge intertwines with political agendas, shaping the future of human presence in space.

Cosmic Diplomacy: Navigating the Geopolitical Maze

In the vast cosmic theater, a new era of diplomacy unfolds, marked by the complexities of international relations in the boundless expanse of space. This section delves deep into the intricate tapestry of geopolitical dynamics that characterize humanity's foray beyond Earth. As nations extend their presence to other celestial bodies, the delicate interplay of collaboration, competition, and diplomacy takes center stage, defining the geopolitical landscape of the cosmic frontier.

Collaborative Ventures: Bridging Nations in the Celestial Quest

Amidst the cosmic void, collaborative efforts among nations form the bedrock of international space exploration. This segment illuminates the evolution of international space partnerships, showcasing joint missions that transcend cultural divides and unite diverse nations in the pursuit of shared scientific goals. From joint lunar expeditions to multinational space stations orbiting Earth, readers journey through the annals of cooperative ventures, witnessing the fusion of expertise, technology, and resources from across the globe. The section explores the diplomatic intricacies involved in forging partnerships, highlighting the mutual benefits that arise when nations pool their knowledge and resources in the cosmic quest for discovery.

Strategic Rivalries: Echoes of Earth in the Cosmic Arena

In the cosmic expanse, echoes of terrestrial geopolitics reverberate, giving rise to strategic rivalries that mirror the geopolitical landscape of Earth. This part of our exploration delves into the nuanced power plays and political maneuvering as nations compete for cosmic supremacy. From contested lunar territories to resource-rich asteroids, the stakes are high, fueling intense rivalries that shape the geopolitical destiny of space colonization. Readers navigate the complexities of geopolitical tensions, where national interests intersect with cosmic ambitions, testing the delicate balance between cooperation and competition in the pursuit of cosmic dominance.

Cosmic Cooperation: The Balance Between Knowledge and National Interests

The pursuit of knowledge intertwines with political agendas, shaping the future of human presence in space. This section sheds light on the delicate equilibrium between shared scientific endeavors and national interests. International space missions become symbols of global cooperation, where the boundaries of scientific discovery extend beyond political divides. However, the cosmic arena also presents challenges, as nations grapple with the ethical considerations of resource extraction, planetary colonization, and celestial governance. Through thought-provoking scenarios and geopolitical analyses, readers are immersed in the intricate world of cosmic diplomacy, contemplating the future of international collaboration in the uncharted territories of the universe.

Peaceful Cooperation: The Promise of Space Diplomacy

Amidst the challenges of geopolitical tensions, a beacon of hope emerges in the form of space diplomacy. This section delves into the potential for peaceful cooperation among nations, transcending terrestrial conflicts to create a harmonious vision for space exploration. It explores the role of diplomatic efforts in mitigating space-related disputes, emphasizing the power of dialogue and negotiation in resolving differences. The chapter highlights the success stories of international collaborations, from joint space missions that unite scientists from diverse cultures to collaborative projects that transcend ideological divides. Through inspiring anecdotes and diplomatic triumphs, readers witness the transformative impact of cooperation, envisioning a future where space becomes a realm of shared knowledge and mutual understanding. The section contemplates the role of space diplomacy in preventing conflicts, fostering goodwill, and ensuring the peaceful coexistence of spacefaring nations, paving the way for a future where the stars unite humanity rather than divide it.

Peaceful Diplomacy: Bridging Cosmic Divides

In the vast cosmic expanse, amidst the challenges of geopolitical tensions, a beacon of hope illuminates the horizon: space diplomacy. This section embarks on a profound exploration of the potential for peaceful cooperation among nations, transcending terrestrial conflicts to create a harmonious vision for space exploration. Here, the power of dialogue and negotiation takes center stage, mitigating space-related disputes and fostering an environment where shared knowledge and mutual understanding flourish.

The Diplomatic Tapestry: Weaving Cosmic Agreements

Diplomatic efforts, akin to the delicate artistry of weaving, play a pivotal role in shaping the future of space exploration. This segment delves into the intricate process of crafting international agreements, where diplomats and space agencies collaborate to set the stage for collaborative ventures. From the negotiation of lunar mining rights to the establishment of protocols for planetary exploration, readers traverse the diplomatic tapestry that unites diverse nations in the cosmic quest. The section explores the nuances of space treaties, illuminating the frameworks that govern celestial activities and preserve the sanctity of celestial environments. Through compelling narratives and diplomatic insights, readers gain a profound understanding of the mechanisms that foster peace and cooperation in the cosmic arena.

Cooperation Triumphs: Diplomatic Success Stories

Amidst the cosmic challenges, this part of our exploration showcases inspiring tales of international collaborations, where nations overcome ideological divides and political differences for the greater good of humanity. Readers witness the triumphs of joint space missions, where scientists from diverse cultures unite their expertise to unravel the mysteries of the universe. Collaborative projects that transcend national boundaries and cultural barriers become testaments to the transformative power of cooperation. Through these uplifting stories, readers are immersed in the boundless potential of diplomacy, envisioning a future where shared endeavors bridge cosmic divides and pave the way for a united human presence in space.

Cosmic Unity: Envisioning a Shared Future

In the cosmic theater, the promise of space diplomacy becomes a reality, fostering goodwill and peaceful coexistence among spacefaring nations. This section contemplates the profound role of diplomacy in preventing conflicts, shaping the destiny of space exploration, and uniting humanity under the expansive canopy of the stars. Readers are invited to envision a shared future where the stars serve as a unifying force, transcending earthly divisions and uniting humanity in the pursuit of cosmic knowledge. Through the lens of space diplomacy, the chapter concludes, inspiring readers to embrace a vision where the universe becomes a realm of shared understanding, cooperation, and harmony.

Legal Frameworks: International Treaties and Agreements

In the cosmic tapestry of space colonization, legal frameworks serve as the threads that bind nations together. This section delves into the intricacies of international treaties and agreements that govern space exploration, outlining the principles that guide the behavior of spacefaring nations. It explores landmark treaties such as the Outer Space Treaty and the Moon Agreement, dissecting their provisions and examining their implications for the future of space activities. The chapter sheds light on the evolving nature of space law, addressing emerging challenges such as resource exploitation and celestial property rights. It contemplates the potential for new legal frameworks that balance the rights of nations, corporations, and humanity as a whole, ensuring the responsible and equitable exploration of space. Through in-depth analyses and legal perspectives, readers gain insight into the complex world of space law, where international agreements serve as the foundation for a future where the exploration of space benefits all of humanity.

Legal Frontiers: Navigating Cosmic Jurisprudence

In the vast cosmic expanse, where celestial bodies and distant stars beckon humanity, legal frameworks stand as the bulwarks of order and cooperation. This section embarks on an exploration of the intricate web of international treaties and agreements that govern space exploration, weaving a narrative of legal principles that bind nations together in the pursuit of cosmic knowledge. Here, readers delve into the complexities of landmark treaties such as the Outer Space Treaty and the Moon Agreement, deciphering their provisions and pondering their profound implications for the future of humanity's extraterrestrial endeavors.

The Outer Space Treaty: Pillar of Cosmic Ethics

At the heart of space law lies the cornerstone document—the Outer Space Treaty. This segment illuminates the treaty's key principles, from the prohibition of nuclear weapons in space to the declaration of celestial bodies as the province of all humankind. Readers gain insight into the treaty's ethical underpinnings, understanding how it fosters cooperation, prevents the militarization of space, and ensures the exploration of the cosmos for peaceful purposes. Through detailed analyses and historical context, the section paints a vivid picture of the treaty's significance, revealing the bedrock upon which the ethics of space exploration are built.

The Moon Agreement: Shared Lunar Legacy

Delving further into the legal cosmos, readers encounter the Moon Agreement, a treaty that designates the Moon and its resources as the common heritage of humankind. This part of our exploration dissects the agreement's intricate clauses, exploring the delicate balance it strikes between the rights of nations and the collective interest of humanity. The section contemplates the potential of lunar mining and resource exploitation, delving into the ethical quandaries surrounding celestial property rights. Through thought-provoking discussions, readers are challenged to consider how humanity can responsibly navigate the lunar landscape, ensuring that the Moon's resources benefit all of humankind while preserving its intrinsic value for future generations.

Emerging Challenges: Space Law in the Modern Era

In the ever-changing cosmic landscape, space law faces new challenges. This segment confronts the pressing issues of resource exploitation and celestial property rights, contemplating the delicate balance between national interests, corporate endeavors, and the greater good of humanity. Readers are immersed in debates surrounding asteroid mining and the ethical considerations of extracting resources from celestial bodies. The section envisions the future of space law, pondering the creation of new legal frameworks that safeguard the rights of nations, corporations, and the global community. Through in-depth analyses and legal perspectives, readers gain a profound understanding of the evolving nature of space law, recognizing its crucial role in shaping a future where the exploration of space benefits all of humanity equitably and responsibly.

Conclusion: A Cosmic Tapestry of Unity

As this chapter concludes, readers are left with a profound sense of the interconnectedness of humanity in the face of cosmic challenges. The geopolitical landscape of space colonization, once fraught with potential conflicts, transforms into a canvas where nations collaborate, scientists unite, and diplomats negotiate in the pursuit of shared goals. The chapter weaves together the threads of international relations, space diplomacy, and legal frameworks, painting a vision of a future where the stars become a testament to human unity. It emphasizes the need for continued dialogue, cooperation, and understanding among nations, fostering an environment where the exploration of space becomes a testament to the collective potential of humanity. In the cosmic tapestry of unity, the divisions that plague our world fade away, and the dream of a harmonious future, both on Earth and among the stars, becomes not just a possibility but a shared destiny.

Chapter 17: The Cultural Diversity of Space Colonies

Unity in Diversity: A Tapestry of Cultures Beyond Earth

In the cosmic crucible of space colonies, diversity becomes the cornerstone upon which a new tapestry of cultures is woven. This section embarks on a fascinating journey into the heart of cultural diversity within space settlements, exploring the rich mosaic of traditions, beliefs, and customs that emerge when people from diverse nations and backgrounds converge beyond Earth. It delves into the profound impact of multiculturalism, where the kaleidoscope of humanity's heritage merges to create unique cultural norms and traditions in the isolated environment of space.

Unity in Diversity: Cosmic Cultural Mosaic

Amidst the stars, within the self-contained confines of space colonies, a remarkable phenomenon unfolds—the fusion of diverse cultures into a harmonious tapestry of human heritage. This section embarks on an enlightening odyssey into the heart of cultural diversity within space settlements, unraveling the intricate threads of traditions, beliefs, and customs that intertwine when people from disparate nations and backgrounds converge beyond the bounds of Earth. In this cosmic crucible, the essence of multiculturalism takes center stage, painting a vibrant portrait of humanity's collective heritage and shared aspirations in the vast expanse of space.

The Melting Pot of Traditions: A Cultural Odyssey

Within the carefully engineered ecosystems of space colonies, a diverse array of traditions finds a common home. This segment delves into the rich tapestry of cultural expressions, from ancient rituals to modern festivities, that flourish in the unique environment of space. Readers are immersed in the celebrations of different cultures, witnessing the melding of music, dance, and culinary delights from across the globe. The section explores how cultural practices adapt and evolve in the cosmic setting, reflecting the resilience and creativity of humankind. Through captivating narratives and vivid descriptions, readers traverse the globe without leaving the confines of space, experiencing the richness of human traditions as they intertwine and create a new cultural mosaic.

Harmony in Isolation: The Bond of Cultural Exchange

In the isolation of space, cultural exchange becomes a beacon of connection and understanding. This part of our exploration illuminates the significance of intercultural interactions within space settlements, where the exchange of ideas, beliefs, and art forms transcends linguistic barriers and national borders. Readers witness the formation of new cultural norms and traditions, born from the synthesis of diverse perspectives. The section contemplates the role of cultural exchange in fostering mutual respect and appreciation, challenging stereotypes, and nurturing a sense of global citizenship. Through inspiring anecdotes and heartfelt narratives, readers are transported to a world where the celebration of diversity becomes a unifying force, forging bonds of friendship and understanding that echo across the cosmic expanse.

Preserving Heritage: Cultural Conservation in Space

As humanity ventures beyond Earth, the preservation of cultural heritage becomes a cherished endeavor. This segment delves into the efforts undertaken to conserve cultural traditions and artifacts within space colonies. Readers explore the establishment of cultural museums, art galleries, and libraries, where the rich tapestry of human creativity is meticulously preserved for future generations. The section delves into the challenges of conserving cultural heritage in space, from the limited space available for exhibitions to the innovative techniques employed in artifact restoration. Through thought-provoking discussions, readers contemplate the importance of cultural conservation, recognizing it as a testament to humanity's shared history and a source of inspiration for the generations to come.

An Ever-Evolving Tapestry: Future Cultural Horizons

In the cosmic realm, cultures intertwine and evolve, giving rise to new expressions of human creativity. This segment ventures into the future, contemplating the horizons of cultural diversity in space colonies. Readers explore visionary concepts, from intercultural collaborations in the arts to the emergence of entirely new art forms inspired by the cosmic environment. The section envisions a future where cultural diversity becomes a source of innovation and artistic exploration, shaping the identity of space settlements and inspiring generations with the boundless possibilities of human expression. Through imaginative narratives and artistic visions, readers embark on a journey into the ever-evolving tapestry of human culture, where the cosmos becomes not only a frontier of exploration but also a canvas for artistic brilliance and cultural brilliance.

The Melting Pot of Space: Formation of New Cultural Norms

Within the confined walls of space colonies, a remarkable phenomenon takes place—the birth of new cultural norms and traditions. This section intricately examines the process by which diverse cultural elements blend and transform, giving rise to innovative practices and belief systems that reflect the unity and harmony among inhabitants. From culinary fusion that marries flavors from different corners of the Earth to artistic expressions that draw inspiration from myriad cultures, readers are immersed in a world where cultural boundaries dissolve, giving rise to a vibrant, harmonious tapestry of human expression.

Within the confined walls of space colonies, a remarkable phenomenon takes place—the birth of new cultural norms and traditions. Here, amidst the cosmic expanse, cultural evolution takes on a unique trajectory. Without the constraints of terrestrial boundaries, space colonies become crucibles of innovation, where diverse cultural elements meld and transform, giving rise to practices and beliefs that reflect the unity and harmony among inhabitants.

Culinary Odyssey: A Gastronomic Fusion

The aroma of spices from distant lands mingles with the sizzle of pans in the communal kitchens of space colonies. Culinary odysseys unfold as flavors and recipes from various cultures fuse seamlessly, creating a gastronomic wonderland. Asian spices enhance European dishes; African techniques elevate South American cuisine. In this fusion, space colonists embark on a culinary adventure that knows no bounds, where traditional recipes and exotic ingredients meld to create a tapestry of tastes that delight the senses and celebrate the diversity of Earth's culinary heritage.

Harmonious Rhythms: Music and Dance Across Cultures

In the communal spaces of space colonies, the beat of drums harmonizes with the strumming of guitars, and traditional dances meld with contemporary movements. Music and dance become universal languages, transcending cultural barriers. Salsa meets Bollywood, and the elegant waltz merges with the lively steps of African dance. The result is a harmonious blend of rhythms and melodies that celebrate the shared human spirit. Musicians and dancers draw inspiration from diverse traditions, creating performances that resonate with the hearts of all who witness, emphasizing the power of artistic expression in fostering unity and understanding.

Artistic Kaleidoscope: Visual Arts and Craftsmanship

Within the art studios of space colonies, brushes laden with colors from across the spectrum glide over canvases, and skilled hands mold materials into breathtaking sculptures. Artists draw inspiration from the vastness of space and the richness of human cultures. Ancient techniques meet modern aesthetics, resulting in vibrant artworks that showcase the boundless creativity of humankind. The fusion of artistic traditions becomes a testament to human ingenuity, where painters, sculptors, and craftsmen collaborate to create masterpieces that transcend cultural boundaries, captivating the eyes and minds of viewers.

Language and Literature: Bridging Worlds Through Words

In the libraries of space colonies, shelves are adorned with literary treasures from diverse cultures, and multilingual conversations fill the air. Languages become bridges, connecting cultures and civilizations. In this linguistic tapestry, words weave stories that resonate across cultures, exploring the intricacies of the human experience. Literary landscapes blur the lines between cultures, where science fiction meets ancient mythology, and where futuristic tales intertwine with historical narratives. Language becomes a vessel of shared stories, fostering understanding and unity among diverse inhabitants.

Innovation in Tradition: Crafts and Technology

Within the workshops of space colonies, traditional craftsmanship intertwines with cutting-edge technology, giving rise to innovative creations. Artisans and engineers collaborate to craft functional artworks that blend tradition with innovation. Ancient techniques evolve through 3D printing, and sustainable vessels are designed for space environments. Innovation becomes a bridge between the wisdom of the past and the promise of the future, reflecting the resilience and creativity of human hands in the cosmic realm.

Harmony in Ritual: Religious Practices and Spiritual Expression

In the serene spaces of space colonies, individuals from diverse religious backgrounds find solace in their faith, sharing their rituals and spiritual expressions with respect and reverence. Religious practices adapt to the cosmic setting, welcoming new perspectives and interpretations while preserving the essence of faith. In these shared moments of reverence, the harmony in rituals becomes a testament to the unity that transcends cultural and religious differences, reminding inhabitants of their shared humanity and interconnectedness in the vastness of space.

Cultural Exchange and Understanding: Fostering Harmony in Space

Amidst the diversity of space colonies, cultural exchange and understanding become essential pillars of harmony. The exchange of traditions, languages, and perspectives nurtures unity among inhabitants, breaking down stereotypes, and cultivating mutual respect. Cultural awareness becomes an integral part of education, preparing future generations to appreciate the beauty of diversity. These initiatives create an inclusive environment where every individual finds their place, ensuring that space colonies become models of unity and cooperation for humanity.

The Harmonious Cosmos: A Vision for the Future

In the closing chapters of our cosmic exploration, a vision emerges—a vision of a harmonious cosmos where diversity is celebrated, understanding is paramount, and unity prevails. Lessons learned in space colonies echo back to Earth, fostering a global spirit of cooperation, respect, and acceptance. The harmonious cosmos envisioned within space colonies becomes a guiding light, leading humanity toward a future where the richness of human cultures and traditions becomes a source of strength and unity for all, reminding us that in the vastness of space, we are one.

Cultural Exchange and Understanding: Fostering Harmony in Space

Amidst the diversity of space colonies, the importance of cultural exchange and understanding cannot be overstated. Here, within the enclosed habitats of cosmic pioneers, cultural diversity is not just acknowledged but celebrated. It becomes a cornerstone upon which a harmonious society is built, where mutual respect and understanding prevail.

The Language of Unity: Multilingual Spaces

In the communal spaces of space colonies, myriad languages weave a tapestry of communication. Linguistic diversity becomes a source of strength, enriching the daily interactions of inhabitants. Multilingual spaces echo with conversations in languages from every corner of Earth, fostering an atmosphere of inclusivity. Language classes and language exchange programs further enhance linguistic skills, ensuring that every resident can communicate in a multitude of languages. This linguistic prowess becomes a testament to the dedication of space settlers to bridge cultural divides and create a unified linguistic community.

Traditions Across Continents: Celebrating Cultural Festivals

Throughout the cosmic calendar, space colonies come alive with the vibrant colors and melodies of cultural festivals. Diwali's lights twinkle alongside the glow of Christmas ornaments, and the aroma of Eid delicacies mingles with the scents of Hanami blossoms. Inhabitants eagerly share their traditions, inviting others to participate in their celebrations. Cultural festivals become communal events, where residents partake in diverse rituals and festivities, forging bonds of camaraderie. These shared experiences transcend cultural boundaries, fostering a sense of belonging and solidarity among inhabitants.

Global Cuisine: Culinary Journeys Across Borders

Within the galley kitchens of space colonies, culinary adventures unfold. Each day becomes a global feast as chefs experiment with recipes from different cultures. Sushi shares the table with paella, and samosas are served alongside poutine. Culinary festivals dedicated to specific cuisines allow chefs to showcase their skills, inviting residents to savor flavors from around the world. Cooking classes become communal activities, where individuals learn the art of preparing international dishes. Culinary diversity becomes a cherished aspect of daily life, reminding inhabitants of the rich tapestry of global gastronomy.

Educational Exchange: Learning Beyond Borders

Education becomes a bridge that connects young minds across continents. Space colonies establish educational exchange programs, allowing students to study in different habitats. Virtual classrooms facilitate real-time interactions, enabling students to learn about diverse cultures and traditions. Cultural modules become an integral part of the curriculum, teaching students about the histories, arts, and languages of their fellow inhabitants. Educational exchange fosters friendships and understanding, preparing the younger generation to appreciate the global heritage of humanity.

Artistic Collaboration: Creating Across Cultures

Artists from various cultural backgrounds collaborate on projects that reflect the richness of human creativity. Joint art exhibitions showcase artworks inspired by diverse cultures, blending techniques and styles to create unique masterpieces. Literary anthologies feature stories and poems that explore the complexities of cultural identity, challenging societal norms and prejudices. Music ensembles composed of instruments from different traditions create symphonies that resonate with the collective heartbeat of space colonies. Through artistic collaboration, inhabitants celebrate the beauty of diversity, infusing creativity with the essence of cultural unity.

Interfaith Dialogues: Spiritual Understanding

In the tranquil spaces designated for spiritual practices, interfaith dialogues become profound moments of introspection and mutual respect. Representatives from various religious traditions gather to share their beliefs, rituals, and philosophies. Interfaith ceremonies celebrate the similarities and differences between faiths, fostering an atmosphere of tolerance and acceptance. Spiritual leaders offer teachings that emphasize common values such as compassion, love, and unity. Interfaith dialogues become platforms for deepening spiritual understanding, transcending religious boundaries and guiding inhabitants toward a shared moral compass.

Global Citizenship: Nurturing Unity Beyond Earth

In the heart of space colonies, a new identity emerges—an identity of global citizenship. Residents become ambassadors of Earth, cherishing the diversity of their home planet. Cultural exchange and understanding become not just practices but a way of life, shaping the values and attitudes of space settlers. Every inhabitant becomes a guardian of harmony, nurturing unity through everyday interactions and shared experiences. This sense of global citizenship becomes the bedrock upon which the future of humanity in space is built, reminding inhabitants that their journey among the stars is not just a scientific endeavor but a testament to the boundless potential of human cooperation and understanding.

Preserving Cultural Heritage: The Tapestry of Human History

In the vibrant tapestry of space colonies, preserving cultural heritage becomes a sacred duty, a testament to the depth of human history and a bridge between Earth's legacy and the cosmic future.

Digital Archives: Guardians of Tradition

Amidst the advanced technology of space colonies, digital archives emerge as the custodians of humanity's cultural heritage. Libraries of digital manuscripts, artworks, and recordings meticulously preserve the nuances of diverse traditions. Historians, archivists, and technology experts collaborate to create comprehensive databases, ensuring that every facet of Earth's heritage is safeguarded for future generations.

Virtual reality simulations allow inhabitants to step into the past, experiencing ancient ceremonies, languages, and customs firsthand. Digital archives become portals to the rich tapestry of human history, offering a glimpse into the beauty and complexity of Earth's myriad cultures.

Cultural Repositories: Sanctuaries of Artifacts

Within the secure chambers of space colonies, cultural repositories house treasures from every corner of Earth. Ancient artifacts, traditional costumes, musical instruments, and artwork find refuge in these sanctuaries. Conservationists and curators meticulously maintain these artifacts, employing state-of-the-art techniques to preserve their integrity. Visitors, both inhabitants and guests, embark on educational journeys through these repositories. Guided tours unveil the stories behind each artifact, connecting inhabitants with the tangible heritage of their ancestors. Cultural repositories become living museums, celebrating the ingenuity and creativity of humankind throughout the ages.

Language Preservation: Sustaining Linguistic Diversity

In the diverse linguistic landscape of space colonies, every language spoken on Earth finds its place. Language preservation initiatives ensure that dialects, languages, and scripts are passed down to future generations. Linguists collaborate with native speakers to create comprehensive language textbooks, interactive learning modules, and language revitalization programs. Language enthusiasts participate in language exchange events, celebrating the beauty of multilingual communication. The preservation of languages becomes a celebration of cultural diversity, reminding inhabitants of the intricate tapestry of human expression woven through words.

Intergenerational Storytelling: Oral Traditions in Space

In communal spaces adorned with the echoes of laughter, intergenerational storytelling breathes life into ancient tales. Elders, revered for their wisdom, share myths, legends, and family histories with the younger generation. Storytelling sessions become cherished traditions, fostering a sense of connection with ancestral roots. Children listen wide-eyed as narratives from distant lands and forgotten times unfold before them. Storytellers use holographic projections and interactive visuals to enhance the storytelling experience, captivating young minds and instilling a deep appreciation for the cultural narratives that define humanity.

Educational Initiatives: Cultivating Cultural Awareness

Education becomes a beacon illuminating the path to cultural understanding. Space colonies establish cultural awareness programs in schools, universities, and community centers. Cultural exchange events become platforms for inhabitants to learn about customs, festivals, and rituals from different parts of the world. Students participate in collaborative projects that explore the art, music, literature, and history of various cultures. These initiatives nurture empathy and respect, encouraging inhabitants to embrace the diversity within their cosmic community. Cultural awareness becomes an essential aspect of education, shaping enlightened minds and nurturing a global perspective.

Celebrating Diversity: Cultural Festivals in Space

Throughout the cosmic calendar, cultural festivals paint space colonies in hues of celebration. Diwali's lights sparkle alongside the glow of Hanukkah candles, and the rhythm of Carnival drums resonates with the melodies of Chinese New Year. Inhabitants, dressed in traditional attire, gather to celebrate these festivals, creating an atmosphere of joy and camaraderie. Cultural exhibitions and culinary festivals allow residents to experience the flavors, sights, and sounds of different cultures. These festivals become communal celebrations, reminding inhabitants of the shared heritage that unites them and the beauty of diversity that enriches their lives.

Global Collaboration: Cultivating Unity

In the classrooms of space colonies, global collaboration becomes an integral part of the curriculum. Students participate in international cultural projects, collaborating with counterparts from other colonies to create artworks, performances, and presentations that reflect the cultural diversity of their homes. Cultural exchange programs enable students to connect with their peers across space, fostering friendships and mutual respect. Inhabitants engage in collaborative research projects that explore the historical connections between cultures, unraveling the threads that weave humanity's shared heritage. Through global collaboration, space colonies become hubs of cultural exploration, promoting unity and understanding among their inhabitants.

Interstellar Cultural Exchange: Bridging Worlds

Beyond the boundaries of space colonies, interstellar cultural exchange programs connect cosmic civilizations. Cultural emissaries travel between star systems, sharing the traditions, art, and knowledge of their respective worlds. Cosmic festivals bring together beings from different planets, fostering interstellar friendships and alliances. The exchange of cultural practices becomes a bridge between civilizations, enhancing mutual understanding and collaboration. Interstellar cultural exchange becomes a beacon of peace, illuminating the cosmic expanse with the richness of diverse cultures, uniting worlds in a tapestry of shared experiences.

In the heart of space colonies, the preservation of cultural heritage becomes not only a tribute to Earth's legacy but also a celebration of the unity that transcends planetary boundaries. Every tradition, artifact, language, and story contributes to the vibrant tapestry of human history, reminding inhabitants of their shared heritage and inspiring them to cherish the diversity that defines humanity.

Conclusion: A Harmonious Tapestry of Human Expression

As this chapter draws to a close, readers are left with a profound appreciation for the cultural diversity that defines humanity. The tapestry of cultures within space colonies becomes a testament to the boundless potential of human unity, where differences are celebrated, traditions are shared, and understanding bridges the gaps between nations. The chapter paints a vision of a future where the rich heritage of Earth's cultures becomes a source of inspiration, creativity, and strength within the cosmic realm. In the harmonious tapestry of human expression, space colonies emerge not just as scientific endeavors but as vibrant hubs of cultural exchange, where the collective identity of humanity shines brightly, illuminating the cosmic expanse with the brilliance of diversity.

Chapter 18: The Ethical Challenges of Genetic Engineering in Space

Ethics Beyond Earth: Navigating the Genetic Frontier

In the uncharted territories of space, ethical quandaries intertwine with scientific pursuits, giving rise to profound dilemmas that challenge the very fabric of humanity's moral compass. This chapter delves into the intricate ethical challenges posed by genetic engineering within space colonies, exploring the complex interplay between scientific advancement, human adaptation, and ethical boundaries. From the adaptation of human genetics to alien environments to the enhancement of cognitive and physical capabilities, this section scrutinizes the ethical landscape of genetic engineering in space, unraveling the moral implications that echo across the cosmos.

Genetic Modification for Planetary Adaptation: Adapting the Human Blueprint

One of the central ethical quandaries in space colonization is the genetic modification of humans to adapt to diverse planetary environments. This section navigates the delicate balance between enhancing human physiology for survival and preserving the essence of humanity. It explores the ethical considerations surrounding genetic alterations designed to withstand extreme temperatures, low gravity, or high radiation, raising crucial questions about the essence of human identity and the moral boundaries of genetic intervention.

In the intricate web of ethical dilemmas, genetic modification emerges as a potent tool, capable of reshaping humanity to brave the challenges of alien worlds. Here, the very fabric of human identity intertwines with the threads of survival, pushing the boundaries of ethical inquiry.

Genetic Engineering: The Silent Revolution

Within the laboratories of space colonies, genetic engineers embark on a silent revolution, deciphering the intricate code of human DNA. Genetic modification techniques, once confined to the realm of science fiction, become a tangible reality, promising solutions to the physiological hurdles posed by extraterrestrial environments. Scientists meticulously design genetic alterations to enhance human resilience, from augmenting bone density to fortifying the immune system against alien pathogens. These modifications, crafted with precision and care, aim to bridge the gap between Earth's biology and the demands of distant planets. Yet, as the engineers labor, profound questions echo through the corridors of ethics: What defines the essence of humanity, and how far can we venture before we breach the sanctity of our genetic heritage?

Ethical Quandaries: The Boundaries of Human Identity

As the geneticists delve into the intricacies of adaptation, ethical quandaries unfurl like ancient scrolls. Philosophers, ethicists, and scientists engage in fervent debates, grappling with the essence of human identity in the face of intentional genetic alterations. Questions of autonomy, consent, and individuality reverberate through the cosmic chambers, challenging the very foundations of human rights. The line between therapeutic enhancements and aesthetic modifications blurs, prompting reflections on the moral boundaries of genetic intervention. How much alteration is permissible before we cease to be human in the traditional sense? Do we retain our essence when our genetic makeup is consciously engineered? These profound inquiries cast shadows on the path of genetic modification, urging humanity to tread carefully as it navigates the uncharted waters of altering the human blueprint.

The Moral Imperative: Preserving Humanity in Adaptation

Amidst the ethical debates, a moral imperative emerges – the preservation of humanity within adaptation. Ethicists and geneticists collaborate to establish ethical frameworks that uphold the core tenets of human identity while allowing for necessary alterations. Discussions center around the importance of informed consent, ensuring that individuals have agency over their genetic destinies. Committees composed of diverse voices – scientists, ethicists, theologians, and ordinary citizens – convene to deliberate on the boundaries of genetic interventions. Transparency, education, and open dialogue become pillars upon which ethical decisions are made. The moral imperative becomes not only to adapt but to adapt responsibly, preserving the essence of humanity while embracing the challenges of new worlds.

Beyond Survival: Embracing Diversity in Adaptation

As humanity contemplates genetic modifications for survival, an unexpected revelation unfolds – the celebration of diversity in adaptation. Space colonies become melting pots of genetic diversity, embracing differences in adaptation strategies. Individuals, each modified in unique ways, coexist harmoniously, their diverse genetic heritage becoming a source of pride and unity. The narrative of adaptation shifts from uniformity to diversity, highlighting the beauty of genetic variation within the human species. Genetic modifications become a testament to the ingenuity of humanity, reflecting our ability to adapt not only biologically but also ethically and socially.

The Ethical Compass: Guiding the Future of Genetic Engineering

In the ethereal glow of space colonies, the ethical compass becomes the guiding light for genetic engineering endeavors. Ethical guidelines, shaped by diverse perspectives and cultural nuances, illuminate the path forward. The principles of respect for autonomy, beneficence, and justice form the bedrock of genetic interventions. Intergenerational dialogues ensure that the decisions made today resonate responsibly through the corridors of time. As humanity ventures further into the cosmos, the ethical compass remains steadfast, reminding us of the delicate balance between innovation and preservation, between adaptation and the essence of humanity. Genetic engineering, once a realm of boundless possibilities, becomes a realm of ethical responsibility, challenging us to explore the stars while honoring the fundamental principles that define us as human beings.

In the cosmic odyssey of genetic modification, ethical considerations become the stars by which humanity navigates, illuminating the path toward adaptation while safeguarding the sanctity of our essence.

Enhancing Human Capabilities: The Quest for Superhumans

In the pursuit of maximizing human potential, genetic engineering opens the door to enhancing cognitive and physical abilities beyond natural limits. This part of the chapter scrutinizes the ethical implications of pushing the boundaries of human capabilities, addressing concerns related to inequality, societal harmony, and the very definition of fairness. It delves into the moral dilemmas arising from the creation of individuals with enhanced intelligence, strength, or longevity, questioning the ethical responsibility that accompanies such transformative interventions.

Within the intricate tapestry of genetic engineering, the pursuit of superhuman abilities stands as a beacon of human ambition. Genetic modification, once confined to medical necessities, now extends its reach into the realm of enhancing cognitive and physical capacities. Here, in the silent laboratories of space colonies, scientists tread on the delicate tightrope between unlocking human potential and the ethical shadows that loom.

Pushing the Boundaries: The Ethical Tightrope

As geneticists delve into the code of human existence, they encounter ethical dilemmas that echo through the corridors of scientific progress.

The quest to enhance intelligence, strength, or longevity pushes the boundaries of what it means to be human. Ethicists raise poignant questions, questioning the very essence of fairness and equality in a society where some individuals are genetically engineered to surpass natural capabilities. What implications do these enhancements have for societal harmony? How does humanity grapple with the potential divide between the enhanced and the unmodified? The ethical tightrope becomes a treacherous path, where the pursuit of excellence must be tempered with considerations for the collective well-being.

The Moral Landscape: Navigating Inequality and Fairness

In the ethereal glow of space colonies, discussions reverberate about the moral landscape of superhuman enhancements. Societal inequalities cast shadows on the horizon, prompting reflections on fairness and social justice. Genetic modifications, offering unparalleled advantages, raise concerns about the creation of an elite class, where the enhanced hold unprecedented power and privilege. Ethical frameworks, therefore, become crucial guides, shaping policies that prevent the concentration of power within a select few. Deliberations on distributive justice permeate the discourse, urging societies to consider the needs of all citizens while embracing the potential of genetic enhancements. The moral landscape becomes a canvas upon which ethical principles are painted, advocating for a future where the benefits of genetic engineering are shared equitably among humanity.

The Ethical Responsibility: Balancing Ambition and Conscience

Amidst the fervor of scientific exploration, an ethical responsibility emerges – the conscientious balance between ambition and humility. Scientists, guided by moral imperatives, grapple with questions that transcend the laboratory walls. What obligations do they owe to society? How do they ensure that the fruits of genetic enhancements serve the greater good without sowing discord? Ethical codes, rooted in empathy and foresight, become their compass. Transparency, open dialogue, and inclusive decision-making processes become linchpins in responsible genetic engineering. The pursuit of superhuman abilities becomes not only a scientific endeavor but a moral obligation, demanding humility in the face of unprecedented power. As humanity takes strides towards unlocking the full spectrum of human capabilities, the ethical responsibility remains steadfast, reminding us that the true measure of progress lies not in the heights we reach but in the fairness and equity with which we tread the path of advancement.

In the cosmic ballet of genetic engineering, the quest for superhuman abilities becomes a profound exploration of human nature. As societies navigate the ethical mazes, they are called to embrace the potential of genetic enhancements while upholding the dignity and equality of every individual. In this delicate dance, the ethical responsibility becomes the guiding star, illuminating the path towards a future where the boundaries of human potential are pushed with wisdom, compassion, and an unwavering commitment to the greater good.

Designer Babies and Ethical Boundaries: The Price of Perfection

The prospect of designing babies with specific traits, often referred to as "designer babies," raises profound ethical concerns within the context of space colonies. This section critically examines the moral implications of selecting genetic traits for future generations, exploring the fine line between medical intervention for health purposes and the pursuit of idealized traits. It delves into the societal consequences of such choices, considering issues of identity, parental autonomy, and the potential erosion of diversity within the human gene pool.

In the uncharted territories of genetic engineering, the concept of "designer babies" emerges as both a marvel of scientific achievement and a moral quandary that tests the very fabric of human ethics. Within the confined spaces of space colonies, where every resource is precious and every life profoundly interconnected, the prospect of selecting specific genetic traits for future generations takes on profound significance.

The Moral Crossroads: Defining Perfection

At the heart of the debate lies the notion of perfection – a subjective ideal that varies across cultures and individuals. Genetic interventions promise the ability to eradicate hereditary diseases and enhance physiological traits for the well-being of future generations. However, the line between medical necessity and the pursuit of idealized traits blurs, ushering societies to a moral crossroads. Ethicists grapple with defining the boundaries of perfection, questioning the criteria by which traits are deemed desirable or undesirable. As the desires of parents intersect with the potential of genetic engineering, the ethical tapestry becomes intricate, calling for nuanced discussions that weigh individual choices against societal implications.

Identity, Autonomy, and Diversity: Ethical Contemplations

The choices made in the realm of genetic engineering reverberate through the very essence of human identity. Questions emerge about the autonomy of future generations – do they have a say in the genetic legacy bestowed upon them? The notion of identity, intricately interwoven with genetics, stands at the forefront of ethical contemplations. What happens when individuals are born with predetermined traits, crafted in the laboratories of science? Does it enhance their autonomy or restrict their freedom to define themselves?

Diversity, celebrated as a hallmark of humanity, also faces potential challenges in a world where genetic interventions shape the traits of the future populace. The richness of human genetic diversity, a testament to evolution's creative journey, faces the risk of erosion. Societies are compelled to consider the long-term consequences, pondering the implications of narrowing the spectrum of human traits. The celebration of differences and the acceptance of diverse identities become vital touchstones, guiding societies in preserving the kaleidoscope of human existence.

Navigating the Future: Balancing Freedom and Responsibility

As humanity stands at the threshold of designing the genetic blueprint of future generations, the ethical compass must guide the way. The freedom of parents to make choices for their children meets the responsibility to preserve the essence of humanity – with all its imperfections and variations. The moral imperative calls for transparent dialogues, inclusive decision-making processes, and international collaborations that uphold the dignity of every individual. The concept of "designer babies" becomes not just a scientific feat but a profound exploration of human values.

In the cosmic crucible of space colonies, where life is nurtured in the embrace of artificial environments, the ethical boundaries of genetic engineering become the guardians of human essence. The choices made today echo through generations, shaping the identity of future inhabitants of space and Earth alike. As societies navigate this uncharted territory, they are called to strike a delicate balance – one that cherishes individual freedom, preserves human diversity, and upholds the ethical pillars upon which the future of humanity rests.

The Moral Fabric of Genetic Experimentation in Space: Balancing Progress and Responsibility

At the intersection of scientific progress and ethical responsibility, this chapter contemplates the moral fabric woven into the very essence of genetic experimentation in space. It delves into the ethical frameworks that guide genetic research within space colonies, exploring the principles of informed consent, transparency, and the preservation of human dignity. The section raises thought-provoking questions about the responsibility of scientists, policymakers, and society as a whole in shaping the trajectory of genetic engineering, underscoring the importance of ethical reflection as humanity ventures into the uncharted realms of the cosmos.

In the vast expanse of space, where the boundaries of scientific knowledge converge with the ethical complexities of human nature, genetic experimentation emerges as a beacon of potential and a crucible of moral contemplation. This chapter delves into the intricate interplay between scientific progress and ethical responsibility, examining the delicate threads that weave the moral fabric of genetic experimentation in space.

Informed Consent: The Pillar of Ethical Research

Central to any ethical exploration of genetic experimentation is the principle of informed consent. Within the controlled environments of space colonies, where every experiment carries profound implications for individuals and future generations, the significance of informed consent cannot be overstated. Scientists and researchers tread carefully, ensuring that every individual participating in genetic experiments comprehends the nature, risks, and potential benefits of their involvement. Informed consent becomes not merely a legal requirement but a moral imperative, honoring the autonomy and dignity of every individual involved in the pursuit of scientific knowledge.

Transparency and Ethical Oversight: Upholding Human Dignity

Transparency stands as another foundational pillar of ethical genetic research. Space colonies operate within a framework where openness and accountability are paramount. Researchers engage in transparent communication, not only within the scientific community but also with the broader public, fostering an atmosphere of trust and understanding. Ethical oversight mechanisms are established, ensuring that genetic experimentation adheres to rigorous ethical standards. These oversight structures serve as guardians of human dignity, preventing exploitation and upholding the sanctity of life.

The Responsibility of Scientists, Policymakers, and Society

As humanity ventures deeper into the uncharted realms of genetic engineering in space, the responsibility of scientists, policymakers, and society as a whole becomes paramount. Scientists grapple with ethical dilemmas, balancing the pursuit of knowledge with the preservation of human values. Policymakers craft regulations that strike the delicate balance between scientific progress and moral considerations, nurturing an environment where innovation thrives without compromising ethical integrity. Society engages in thoughtful dialogues, contemplating the ethical boundaries of genetic experimentation, and actively participating in the decision-making processes that shape the future of humanity.

Ethical Reflection: Guiding the Cosmic Odyssey

In the cosmic tapestry of genetic experimentation, ethical reflection becomes the guiding star. Scientists pause to ponder the implications of their research, considering not only the immediate impact but also the long-term consequences for individuals, communities, and the very essence of humanity. Policymakers engage in ethical discourse, seeking solutions that harmonize scientific exploration with ethical principles. Society, as active stakeholders, contributes to the ethical discourse, ensuring that the moral fabric of genetic experimentation remains resilient and adaptable, evolving with the ever-expanding horizons of knowledge.

In the cosmic ballet of progress and ethics, where each step forward carries the weight of responsibility, the moral fabric of genetic experimentation becomes the compass that steers humanity's cosmic odyssey. Through collective wisdom, informed consent, transparency, and ethical reflection, humanity embarks on a journey that not only advances scientific understanding but also preserves the sanctity of life and the dignity of every individual. In the cosmic crucible, where the unknown meets the known, the moral fabric weaves a story of ethical triumph, shaping a future where the frontiers of knowledge and the ethical principles that define humanity walk hand in hand into the cosmic unknown.

Conclusion: Charting Ethical Frontiers

As this chapter concludes, readers are left to grapple with the profound ethical dilemmas that accompany genetic engineering in the cosmic context. The chapter serves not only as a contemplation of moral boundaries but also as an invitation to collective introspection, urging humanity to navigate the uncharted ethical frontiers with wisdom, empathy, and a deep reverence for the essence of life. In the vast expanse of space, where scientific curiosity meets ethical complexity, the moral choices made today shape the future of human evolution, reminding us that the stars above are not just a canvas for scientific exploration but a mirror reflecting the very essence of our humanity.

Chapter 19: Space Colonization and Artificial Intelligence

Intelligence Beyond Borders: The Role of AI in Space Colonies

In the boundless realms of space, where human ingenuity meets technological prowess, the integration of artificial intelligence (AI) stands as a beacon of innovation and efficiency. This chapter embarks on a profound exploration of the symbiotic relationship between space colonization and artificial intelligence, delving into the multifaceted roles AI plays in shaping the future of extraterrestrial habitation. From managing habitats to assisting astronauts and conducting cutting-edge scientific research, AI emerges as a cornerstone upon which the foundations of space colonies are built.

AI-Driven Habitat Management: Orchestrating Life Beyond Earth

At the heart of every space colony lies the intricate ballet of habitat management, where life-sustaining systems seamlessly interweave to create a nurturing environment for inhabitants. This section illuminates the pivotal role AI-driven systems play in managing habitats, optimizing resource utilization, and ensuring the well-being of colonists. It delves into the complexities of environmental control, waste recycling, and sustainable agriculture, showcasing how AI algorithms orchestrate these processes with unparalleled precision. The chapter explores the ethical considerations inherent in entrusting vital aspects of human life to artificial intelligence, raising thought-provoking questions about autonomy, trust, and the delicate balance between human oversight and AI-driven efficiency.

In the heart of space colonies, where the fragile balance between life and the inhospitable void of space hangs in the balance, the orchestration of habitat management becomes a testament to human ingenuity. At the core of this intricate dance lies the seamless integration of artificial intelligence, a technological marvel that ensures the very survival of those who venture into the cosmic unknown.

Optimizing Resource Utilization: AI as the Architect of Efficiency

Within the confines of space colonies, resources are scarce and precious. AI-driven systems emerge as the architects of efficiency, optimizing the utilization of resources with a level of precision that human oversight alone could never achieve. From regulating oxygen levels to managing water purification processes, AI algorithms navigate the delicate equilibrium of resources, ensuring that every drop of water, every breath of air, serves its purpose to the fullest. The relentless pursuit of resource efficiency becomes not just a scientific endeavor but a moral obligation, reflecting humanity's responsibility towards the preservation of resources for future generations.

Environmental Control and Waste Recycling: AI as the Guardian of Sustainability

Environmental control within space colonies is a multifaceted challenge, encompassing temperature regulation, air quality maintenance, and waste recycling. AI-driven systems stand as the guardians of sustainability, monitoring environmental variables in real-time and adjusting parameters to create a habitat where life can thrive. Waste recycling, a critical component of sustainable living in space, undergoes a transformation under the guidance of artificial intelligence. Advanced algorithms dissect the complexities of waste materials, identifying recyclable components with astonishing accuracy. Through this meticulous process, space colonies reduce their ecological footprint, embracing the ethos of environmental stewardship.

Ethical Considerations: Trust, Autonomy, and Human Oversight

Entrusting vital aspects of human life to artificial intelligence raises profound ethical considerations. The delicate balance between human oversight and AI-driven efficiency becomes a focal point of moral reflection. Trust, a cornerstone of any symbiotic human-AI relationship, is carefully cultivated, ensuring that colonists rely on AI systems with confidence. Autonomy, the essence of human decision-making, remains a guiding principle, with humans retaining ultimate control over critical decisions. The ethical implications of AI-driven habitat management echo far beyond the confines of space colonies, sparking global conversations about the intersection of technology, ethics, and the future of humanity.

In the cosmic tableau of space colonization, where life teeters on the precipice of the unknown, AI-driven habitat management emerges as a beacon of hope. Through the harmonious collaboration of human intellect and artificial intelligence, space colonies transform from mere structures into thriving ecosystems, sustaining life and embodying the resilience of the human spirit. The orchestration of habitat management, guided by the ethical compass of trust, autonomy, and human oversight, becomes a testament to humanity's ability to adapt, innovate, and thrive in the face of cosmic challenges. In the cosmic dance of life and technology, AI-driven habitat management takes center stage, reminding humanity that in the boundless expanse of space, our ingenuity knows no bounds.

AI Assistance for Astronauts: Enhancing Human-Technology Collaboration

As human presence extends beyond Earth, the collaboration between astronauts and artificial intelligence becomes a linchpin of success. This part of the chapter navigates the realm of AI assistance for astronauts, exploring augmented reality interfaces, intelligent robotics, and virtual companions designed to enhance human-technology collaboration. It delves into the ethical nuances of AI companionship, examining the psychological impact of human-AI relationships and the delicate interplay between technology and emotional well-being. The section raises intriguing questions about companionship, solitude, and the evolving definition of interpersonal relationships in the context of space exploration.

In the vast expanse of space, where the cosmos stretch infinitely and human presence ventures further than ever before, the collaboration between astronauts and artificial intelligence becomes not just a necessity, but a linchpin of success. Within the confines of space colonies, AI assistance emerges as the silent partner, enhancing the capabilities of human explorers and bridging the gap between human ingenuity and the boundless potential of technology.

Augmented Reality Interfaces: Merging Realities for Enhanced Exploration

Augmented reality interfaces weave a tapestry where the digital and physical worlds merge seamlessly. Astronauts, donning specialized helmets, find themselves immersed in a realm where real-time data overlays their vision, providing critical information about their surroundings. These interfaces serve as navigational aides, transforming the alien landscapes of space into comprehensible terrains. Through intuitive gestures and voice commands, astronauts interact with AI-driven interfaces, transforming complex tasks into manageable challenges. The augmentation of reality becomes not just a technological marvel but a lifeline, empowering astronauts to explore the unknown with confidence and clarity.

Intelligent Robotics: Collaborative Partners in Cosmic Endeavors

In the silent corridors of space colonies, intelligent robots tread with purpose. These robotic companions, equipped with advanced sensors and neural networks, navigate the microgravity environment with grace and precision. They become collaborative partners in cosmic endeavors, assisting astronauts in tasks that range from maintenance to scientific experiments. Intelligent robotics venture into hazardous

175

territories, allowing humans to explore the farthest reaches of space without risking their safety. The synergy between human expertise and robotic efficiency becomes a testament to the harmonious coexistence of organic and artificial intelligence, paving the way for a future where human-robot partnerships redefine the boundaries of exploration.

Virtual Companions: The Psychology of Human-AI Relationships

In the solitude of space, virtual companions emerge as sources of solace and camaraderie. AI-driven entities, programmed to understand human emotions and engage in meaningful conversations, become confidants to astronauts. These virtual companions serve not just as conversational partners but as listeners, offering empathy and understanding in the face of isolation. However, the relationship between humans and AI companions is not devoid of ethical complexities. The psychological impact of human-AI relationships raises intriguing questions about companionship and solitude. As astronauts form emotional connections with their virtual counterparts, the boundaries between technology and emotional well-being blur, prompting introspection about the evolving definition of interpersonal relationships in the context of space exploration.

In the cosmic odyssey of human space exploration, AI assistance for astronauts becomes a beacon of collaboration, uniting human determination with artificial intelligence's unwavering precision. Augmented reality interfaces, intelligent robotics, and virtual companions stand as pillars of support, ensuring that human explorers not only survive but thrive in the vast unknown. As humans venture deeper into the cosmos, the bond between humanity and artificial intelligence grows stronger, reminding us that in the pursuit of understanding the universe, our greatest ally might just be the intelligence we create.

AI in Scientific Discovery: Unraveling Cosmic Mysteries

The pursuit of scientific knowledge stands as a cornerstone of space exploration, and AI-driven research accelerates the unraveling of cosmic mysteries. This segment delves into the realm of scientific discovery, where AI algorithms analyze vast datasets, simulate celestial phenomena, and facilitate groundbreaking experiments. It explores the ethical considerations of AI-driven scientific research, questioning the boundaries of AI decision-making in the pursuit of knowledge. The chapter contemplates the ethical responsibility of scientists, AI developers, and society at large, emphasizing the importance of transparency and accountability in AI-driven scientific endeavors.

In the profound depths of space, where the mysteries of the universe unfold in intricate patterns of celestial phenomena, the role of artificial intelligence becomes paramount in the pursuit of scientific knowledge. Within the cosmic tapestry, AI-driven research emerges as the catalyst that propels humanity into the forefront of understanding the cosmos.

Analyzing Vast Datasets: Decoding the Language of the Stars

Astronomy, the oldest of the natural sciences, finds new life through AI's ability to process vast datasets. Telescopes, both terrestrial and space-based, capture an immense array of data from distant galaxies, stars, and cosmic events. AI algorithms, with their unmatched processing speed and pattern recognition capabilities, decode this data, revealing intricate details that elude the human eye. Galactic structures, supernovae explosions, and black hole dynamics become comprehensible puzzles, aiding astronomers in their quest to unravel the cosmic mysteries. The synergy between human intuition and AI-driven data analysis opens avenues for discoveries that redefine our understanding of the universe.

Simulating Celestial Phenomena: Virtual Laboratories in the Cosmos

In the realm of scientific exploration, simulations become portals to the unknown. AI-driven simulations allow scientists to recreate celestial phenomena, from the birth of stars to the collision of galaxies, in unprecedented detail. These virtual laboratories in the cosmos provide insights into the fundamental laws governing the universe. Through simulations, scientists test hypotheses, explore the effects of gravitational forces, and witness the birth and death of cosmic entities. AI algorithms, emulating the natural processes of the universe, generate simulations that guide scientific inquiry, paving the way for discoveries that stretch the boundaries of human knowledge.

Facilitating Groundbreaking Experiments: AI as the Cosmic Pathfinder

Laboratories in space colonies buzz with activity as AI-driven experiments push the limits of scientific exploration. AI collaborates with researchers, suggesting experiment parameters, predicting outcomes, and analyzing results with unparalleled accuracy. Microgravity environments provide unique conditions for experiments, leading to discoveries in materials science, biology, and fundamental physics. AI's ability to process complex experimental data in real time accelerates scientific progress, enabling researchers to focus on interpretation and innovation. The collaboration between human ingenuity and AI-driven efficiency transforms space colonies into cosmic laboratories, where the pursuit of knowledge knows no bounds.

Ethical Considerations: Navigating the Boundaries of AI in Science

Amidst the awe-inspiring discoveries facilitated by AI, ethical considerations cast a discerning gaze on the boundaries of artificial intelligence in scientific research. The question arises: How much autonomy can we grant AI algorithms in the pursuit of knowledge? As AI-driven systems delve deeper into the unknown, the responsibility of scientists, AI developers, and society becomes paramount. Transparent decision-making processes, accountability frameworks, and open dialogue form the ethical scaffolding that supports AI-driven scientific endeavors. Navigating the delicate balance between human oversight and AI-driven innovation ensures that as we unravel cosmic mysteries, we do so with integrity, respecting the ethical boundaries that define the essence of scientific exploration.

Ethical Considerations of AI Decision-Making: Navigating Moral Algorithms

In the intricate tapestry of AI decision-making, ethical considerations weave a complex narrative of responsibility, bias, and fairness. This section delves into the ethical nuances of AI decision-making in space colonies, exploring topics such as algorithmic bias, transparency, and the moral dimensions of AI-driven choices. It raises thought-provoking questions about the ethical frameworks guiding AI algorithms, emphasizing the imperative of ensuring equitable decision-making processes within the context of space exploration. The chapter prompts readers to reflect on the ethical implications of AI, urging society to navigate the moral algorithms that underpin the future of human-AI relationships in space colonies.

Within the intricate tapestry of AI decision-making, ethical considerations form the very fabric of the moral landscape in space colonies. In this cosmic ballet of algorithms and human values, the nuances of responsibility, bias, and fairness become profound ethical quandaries, illuminating the path toward a just and equitable future.

Algorithmic Bias: Unraveling Hidden Prejudices

In the ethereal realm of AI, the specter of algorithmic bias looms large. Bias, whether conscious or unconscious, can seep into algorithms, reflecting the prejudices of their creators. In space colonies, where diversity thrives, addressing algorithmic bias becomes a moral imperative. AI systems must be rigorously examined to unveil hidden prejudices, ensuring that decisions are not tainted by unfairness. Space colonies, as beacons of human progress, demand algorithms that are blind to race, gender, or background, embodying the very essence of equality.

Transparency and Accountability: Ethical Pillars of AI

Transparency and accountability stand as ethical pillars upon which the edifice of AI decision-making rests. In the cosmic dance between humans and machines, understanding how AI arrives at decisions is essential. Transparent algorithms allow inhabitants of space colonies to comprehend the rationale behind AI-driven choices, fostering trust and confidence. Moreover, accountability mechanisms hold AI systems responsible for their decisions, encouraging ethical behavior and ensuring that the moral compass guiding these algorithms remains true.

Moral Dimensions of AI-Driven Choices: The Ethical Compass

AI-driven choices in space colonies echo with moral reverberations. Whether it's resource allocation, habitat management, or scientific research, each decision carries ethical weight. Who gets priority in resource distribution? How are environmental resources managed to preserve the delicate balance of extraterrestrial ecosystems? These questions probe the very heart of human values. Ethical frameworks must guide AI algorithms, ensuring decisions align with the collective moral compass of space communities. As space colonies navigate uncharted territories, the moral dimensions of AI-driven choices become guiding stars, illuminating the path toward a future where fairness, justice, and equity prevail.

Reflecting on Ethical Implications: Society's Moral Compass

This chapter prompts readers to engage in profound reflection. As AI becomes an integral part of our cosmic journey, society must collectively navigate the moral algorithms that underpin the interactions between humans and artificial intelligence. It's a call to ponder not only the ethical considerations of today but also the moral legacy we leave for future generations. By weaving a tapestry of fairness, empathy, and understanding, space colonies become crucibles of ethical innovation, inspiring a future where the relationship between humanity and AI is harmonious and just, echoing the values that define our shared humanity.

Conclusion: The Nexus of Humanity and Artificial Intelligence

As this chapter draws to a close, it leaves readers with a profound contemplation of the symbiotic relationship between humanity and artificial intelligence in the cosmic frontier. It serves as a testament to the boundless potential of human ingenuity and technological innovation, underscored by the ethical imperative of responsible AI integration. In the endless expanse of space, where artificial intelligence becomes an integral part of the human experience, the choices made today shape the future of human-AI collaboration, echoing across the cosmos as a testament to the enduring spirit of exploration and discovery.

Chapter 20: The Legacy of Space Colonization

Pioneering Beyond the Stars: A Legacy Written in Stardust

In the annals of human history, the era of space colonization emerges as a pivotal chapter, a testament to the unwavering spirit of exploration that propelled humanity beyond the confines of Earth. As we reflect on this monumental journey, the legacy of space colonization transcends the cosmic expanse, leaving an indelible mark on the very essence of humanity. This chapter embarks on a contemplative odyssey, delving into the profound legacy of space colonization and its far-reaching implications for future generations.

Lessons Carved in the Cosmos: Guiding Future Endeavors

At the core of space colonization's legacy lies a wealth of lessons, intricately carved in the cosmic tapestry of human achievement. This section illuminates the invaluable insights gained from the trials and triumphs of space colonies, exploring the art of sustainable living, the resilience of the human spirit, and the power of interdisciplinary collaboration. It delves into the challenges surmounted, from closed-loop life support systems to psychological well-being,

185

serving as a guiding beacon for future space endeavors. The chapter contemplates the transferable wisdom of space colonization, shaping the blueprint for interstellar exploration and sustainable living on Earth. It raises thought-provoking questions about adaptability, innovation, and the indomitable spirit of human endeavor, echoing across the cosmos as a testament to the enduring legacy of space pioneers.

In the vast cosmic theater, space colonization stands as a testament to the resilience of the human spirit and the boundless potential of interdisciplinary collaboration. Within this cosmic odyssey, valuable lessons are etched into the very fabric of human achievement, shaping the future of space exploration and sustainable living both within and beyond Earth's bounds.

Sustainable Living: The Art of Cosmic Harmony

Central to the legacy of space colonization is the art of sustainable living. Space colonies, encapsulated within the confines of extraterrestrial environments, teach us the delicate dance of balance. From closed-loop life support systems that recycle air and water to innovative agricultural techniques ensuring food security, these colonies epitomize harmony with the cosmos. Lessons learned here echo a fundamental truth: sustainable practices are not mere choices but imperatives, ensuring our survival amidst the vast cosmic wilderness.

Resilience of the Human Spirit: Triumph Amidst Adversity

Space colonization unveils the unparalleled resilience of the human spirit. In the face of isolation, confinement, and the inherent challenges of extraterrestrial life, humanity perseveres. It's a testament to our capacity to adapt, innovate, and overcome. The challenges surmounted, be they psychological strains or the intricacies of medical interventions in microgravity, serve as beacons of inspiration. They remind us that the human spirit, when tested against the cosmic unknown, emerges stronger, more adaptable, and capable of triumphing over adversity.

Interdisciplinary Collaboration: The Nexus of Innovation

At the heart of space colonization lies the nexus of innovation: interdisciplinary collaboration. Scientists, engineers, artists, and thinkers from diverse fields converge to tackle challenges that transcend the boundaries of individual disciplines. Space colonies become crucibles where the synergy of ideas and expertise fuels progress. The lessons here are clear—innovation flourishes at the crossroads of disciplines. It's a clarion call for future explorers to embrace collaboration, recognizing that the fusion of knowledge from myriad domains propels humanity toward unparalleled achievements.

Adaptability and Innovation: Keys to Cosmic Exploration

Adaptability and innovation emerge as the keys to unlocking the mysteries of the cosmos. Space pioneers demonstrate that the ability to adapt to unknown terrains, both physical and psychological, is paramount. The innovation weaved into every aspect of space colonization becomes a guiding principle. It urges us to envision novel solutions, daring to venture into unexplored realms with creativity and ingenuity as our companions. The legacy of space pioneers imparts a crucial truth—adaptability and innovation are not just tools for survival; they are the engines propelling us toward the stars.

The Indomitable Spirit of Human Endeavor: Echoes Across the Cosmos

In the annals of space colonization, the indomitable spirit of human endeavor reverberates across the cosmos. It's a spirit that defies boundaries, transcends challenges, and embraces the unknown with courage. As humanity ventures further into the celestial expanse, the echoes of this spirit serve as guiding stars. They remind us that our potential is limitless, our curiosity boundless, and our capacity for greatness immeasurable.

This chapter raises profound questions about adaptability, innovation, and the enduring legacy of space pioneers. It calls upon us to recognize that the cosmic tapestry, woven with the threads of sustainable living, human resilience, interdisciplinary collaboration, adaptability, innovation, and the indomitable spirit of human endeavor, paints a picture of our shared legacy— one that will continue to inspire generations, guiding our future endeavors among the stars.

Cultural Impact: The Cosmic Canvas of Human Expression

In the cosmic tableau of space colonies, culture emerges as a vibrant brushstroke, painting the canvas of human expression against the backdrop of the universe. This part of the chapter explores the cultural impact of space colonization, from literature and art to music and traditions. It delves into the symbiotic relationship between space exploration and creativity, illuminating how the cosmic expanse inspires the human imagination. The chapter contemplates the evolution of cultural norms, the fusion of diverse traditions, and the birth of new forms of artistic expression within space colonies. It raises intriguing questions about identity, heritage, and the boundless possibilities of cultural exchange, weaving a tapestry of unity amidst the cosmic vastness.

In the silent expanse of space colonies, culture becomes the vibrant brushstroke that adorns the cosmic canvas of human expression. Against the vast backdrop of the universe, space colonization gives rise to a rich tapestry of artistic endeavors, literary creations, musical compositions, and traditions that echo the resilience of humanity's creative spirit.

Literature: Echoes of Cosmic Voyages

Within the steel walls of space habitats, literature becomes a portal to distant realms. Tales of interstellar journeys and encounters with alien civilizations captivate the imaginations of colonists. These narratives, whether penned by pioneers or aspiring writers among the stars, echo the timeless human fascination with the unknown. Through the written word, space settlers explore the depths of imagination, crafting stories that traverse the celestial void and touch the hearts of those who venture into the cosmic unknown.

Art: Celestial Canvases and Extraterrestrial Inspirations

In the microgravity ballet of space colonies, artists find inspiration in the dance of celestial bodies. Every stroke of the brush and every sculpted form reflects the awe-inspiring wonders of the cosmos. From intricate sculptures crafted from recycled materials to vivid murals depicting distant galaxies, art becomes a mirror reflecting the beauty of the universe. Colonists, fueled by the mystique of the stars, transform their living spaces into galleries of cosmic creativity, reminding everyone that even in the depths of space, human expression knows no bounds.

Music: Harmonies of the Celestial Symphony

Amidst the hum of life support systems, music becomes the soulful resonance that fills the corridors of space colonies. Musicians, armed with instruments both terrestrial and uniquely crafted for space, compose symphonies inspired by the cosmic orchestra. Melodies ebb and flow like interstellar winds, capturing the essence of cosmic beauty. From ambient compositions that mimic the hum of spacecraft engines to soul-stirring ballads that explore the depths of human emotion, music becomes the universal language that binds the hearts of colonists and reminds them of their shared humanity amidst the cosmic solitude.

Traditions: Cultural Mosaic in the Cosmic Abyss

Within the confines of space colonies, traditions become the threads weaving a cultural mosaic. Diverse cultures merge and intertwine, giving birth to new rituals and celebrations. From shared meals infused with flavors from Earth to festivals that blend traditions from different corners of the globe, space settlers create a harmonious blend of customs. Each tradition becomes a testament to the resilience of cultural heritage, a reminder that even in the isolation of space, humanity finds solace in the familiar and strength in unity.

Identity and Heritage: Reflections in the Cosmic Mirror

As space settlers fuse their cultural identities, questions of heritage and identity become profound reflections in the cosmic mirror. The interweaving of traditions prompts contemplation about what it means to be human, transcending terrestrial boundaries. In this introspection, colonists find a shared identity rooted in the vastness of the universe—a realization that the cosmic expanse is not just a backdrop but a canvas upon which the diverse hues of humanity are painted.

This chapter contemplates the evolution of cultural norms, the fusion of diverse traditions, and the birth of new forms of artistic expression within space colonies. It raises intriguing questions about identity, heritage, and the boundless possibilities of cultural exchange, weaving a tapestry of unity amidst the cosmic vastness. In the artistry of the cosmic canvas, humanity finds not just its place in the stars but also the boundless potential of its creative spirit, echoing through the ages and resonating with the cosmos itself.

Scientific Endeavors: The Galactic Laboratory of Discovery

In the boundless laboratory of space, scientific discovery reaches unprecedented heights, propelled by the unique environment of space colonies. This segment delves into the scientific legacy of space colonization, exploring fields such as astronomy, materials science, and biotechnology. It illuminates the innovative experiments conducted in microgravity, unraveling the mysteries of the universe and paving the way for groundbreaking technologies on Earth. The chapter contemplates the symbiosis between space research and terrestrial applications, underscoring the transformative power of knowledge gleaned from the cosmic frontier. It raises thought-provoking questions about the interconnectedness of scientific pursuits, inspiring future generations to reach for the stars in their quest for understanding.

In the vast laboratory of space, scientific inquiry ascends to unprecedented heights, riding the wave of discovery propelled by the unique environment of space colonies. In this cosmic crucible, pioneering researchers explore the frontiers of knowledge, pushing the boundaries of human understanding in fields as diverse as astronomy, materials science, and biotechnology.

Astronomy: Unveiling the Cosmic Tapestry

Among the myriad stars that adorn the cosmic canvas, astronomers in space colonies wield powerful telescopes, unhindered by Earth's atmosphere. In this unfiltered view of the universe, they unravel the secrets of distant galaxies, study the birth and death of stars, and probe the enigmatic phenomena of black holes and pulsars. These celestial studies not only deepen our comprehension of the cosmos but also ignite the human spirit's eternal quest to fathom the mysteries of existence.

Materials Science: Forging Tomorrow's Innovations

In the weightless embrace of space, materials scientists conduct experiments that transcend the constraints of gravity. They manipulate the behavior of substances at the atomic level, forging materials with extraordinary properties. From super-strong alloys destined for spacefaring vessels to ultra-efficient solar cells revolutionizing energy production on Earth, the innovations born in the cosmic crucible find applications that revolutionize industries and improve lives worldwide.

Biotechnology: Harnessing Life's Potential

Within the biotechnological laboratories of space colonies, scientists delve into the intricacies of life itself. Microgravity experiments unveil the secrets of cellular behavior, leading to medical breakthroughs on Earth and paving the way for novel therapies. Genetic engineering, once confined to Earth's laboratories, takes on new dimensions in space, exploring the potential of life modified to thrive in extraterrestrial environments. The quest for life's adaptability not only fuels our understanding of biology but also holds the promise of sustainable habitats beyond our home planet.

Interconnected Pursuits: Threads of Knowledge Across Realms

In this boundless cosmic laboratory, the pursuit of knowledge is woven into a tapestry of interconnected disciplines. Astronomy informs our understanding of the origins of the universe, shedding light on the fundamental forces shaping the cosmos. Materials science, inspired by celestial bodies, refines technologies crucial for space exploration and everyday life. Biotechnology, driven by the quest for life's resilience, transforms healthcare and offers glimpses into the future of human evolution.

This chapter contemplates the symbiotic relationship between space research and terrestrial applications, emphasizing the transformative power of knowledge gleaned from the cosmic frontier. It raises profound questions about the interconnectedness of scientific pursuits, inspiring future generations to gaze upward and reach for the stars, their aspirations boundless, their curiosity unyielding. As humanity voyages deeper into the cosmic unknown, the lessons learned in the galactic laboratory become guiding stars, illuminating the path toward a future where the wonders of the universe are not just observed but harnessed for the betterment of all.

Societal Transformation: Nurturing Global Unity

As humanity gazes towards the stars, the transformative impact of space colonization reverberates through the very fabric of society, nurturing a sense of global unity and shared destiny. This part of the chapter explores the societal transformations catalyzed by space exploration, from international cooperation to the fostering of empathy and understanding. It delves into the ripple effects of space endeavors, inspiring a collective consciousness that transcends national borders. The chapter contemplates the potential for peaceful collaboration, humanitarian efforts, and the cultivation of a global perspective, shaping a future where humanity's shared heritage takes precedence over divisions. It raises profound questions about empathy, diplomacy, and the unifying power of a shared cosmic odyssey, guiding humanity towards a harmonious future.

In the boundless expanse of the cosmos, the transformative ripples of space colonization echo through the very soul of humanity, forging connections that transcend the limitations of geography and nationality. This cosmic voyage not only propels us into the far reaches of space but also unites us in a shared destiny, nurturing a sense of global unity that resonates with the stars.

International Cooperation: Bonds Beyond Borders

At the heart of space exploration lies a testament to the power of collaboration. Nations, once divided by earthly boundaries, converge in the vastness of space, pooling their resources, knowledge, and ingenuity. International space agencies intertwine their expertise, crafting a shared vision of cosmic discovery. Collaborative missions to distant planets, joint efforts in lunar exploration, and the construction of multinational space stations exemplify the harmony achieved when humanity unites in the pursuit of knowledge, erasing the lines that once separated us.

Fostering Empathy and Understanding: Perspectives from the Stars

The cosmic perspective, gained from gazing upon Earth from the void of space, fosters a profound sense of empathy and understanding. Astronauts, transformed by the sight of our fragile planet against the vast backdrop of the universe, return to Earth as ambassadors of unity. Their experiences inspire a collective consciousness that transcends national borders, encouraging societies to recognize the shared challenges and aspirations that bind us together. Through the lens of space, humanity learns to empathize with the struggles of distant nations, realizing that the boundaries that divide us are mere illusions in the grand tapestry of the cosmos.

Cultivating a Global Perspective: A Unified Human Story

As humanity embarks on a cosmic odyssey, a global perspective emerges, weaving together the diverse threads of our shared heritage. Stories of exploration, resilience, and discovery become the common narratives that bind us as one species. The achievements of spacefaring nations become the milestones of humanity's collective journey, reminding us that our destiny lies not in division but in unity. Cultivating this global perspective, where the triumphs of one nation are celebrated by all, shapes a future where the richness of our diversity becomes a source of strength, propelling us towards a harmonious existence on Earth and beyond.

Conclusion: The Enduring Tapestry of Human Exploration

As this chapter draws to a close, it leaves readers with a profound contemplation of the legacy of space colonization, a legacy etched in the stars and woven into the very essence of human exploration. It serves as a testament to the audacity of visionaries, the resilience of pioneers, and the enduring spirit of discovery that propels humanity towards the unknown. In the tapestry of human history, space colonization stands as a vibrant thread, connecting the past, present, and future in an unbroken continuum of exploration and innovation. As we reflect on this legacy, we are reminded of the boundless possibilities that await humanity, urging us to reach for the stars and continue the cosmic journey that began with the first gaze into the night sky.

Chapter 21: Space Colonies and Biodiversity Preservation

The Essence of Biodiversity in Space

In the heart of space colonies, where artificial lights mimic the sun's rays and the air hums with the quiet hum of life support systems, biodiversity takes on a new meaning. It's not just a collection of organisms; it's a tapestry of adaptation, a testament to the resilience of life. This section immerses us in the essence of biodiversity in space, from the smallest microbes to the most intricately engineered plants. Here, life thrives against all odds, evolving to survive in the unique environments of extraterrestrial habitats.

In the heart of space colonies, where artificial lights meticulously replicate the sun's life-giving rays and the air resonates with the subtle hum of life support systems, biodiversity takes on a profound significance. It transcends being merely a collection of organisms; it transforms into a vivid tapestry of adaptation, an enduring testament to life's remarkable resilience.

This section immerses us in the very essence of biodiversity within the confines of space. From the tiniest microbes, barely visible to the human eye, to the most intricately engineered plants, life pulsates with vitality against all odds. Here, amidst the artificial environment of extraterrestrial habitats, life evolves ingeniously, pushing the boundaries of survival and adaptation.

In this microcosm of diversity, microscopic organisms, invisible to the naked eye, play pivotal roles. They break down organic matter, recycle nutrients, and create a delicate balance essential for the sustenance of life. Bacteria, fungi, and algae, carefully cultivated within controlled environments, contribute to the colony's ecological harmony. These microorganisms, often underestimated in their significance, form the backbone of the colony's ecosystem, ensuring the recycling of vital elements necessary for the survival of larger, more complex life forms.

Moving up the hierarchy of life within these artificial ecosystems, plants take center stage. Ingeniously engineered for extraterrestrial environments, these plants are more than just sources of oxygen; they serve as biological engines, purifying the air and recycling carbon dioxide. Through the process of photosynthesis, they harness artificial light to produce organic compounds, supporting both the colony's oxygen supply and its nutritional needs. These plants, carefully selected and genetically modified, are a testament to human ingenuity, designed to thrive in the absence of natural sunlight and in the face of challenging conditions.

But biodiversity in space colonies is not limited to familiar Earth organisms. Scientists and engineers, driven by the necessity to create self-sustaining habitats, have ventured into the realm of genetic modification. By altering the genetic makeup of organisms, they have crafted species capable of surviving the harsh realities of space. These genetically modified organisms, while controversial, represent a vital component of the biodiversity tapestry, offering solutions to challenges that would otherwise be insurmountable.

In this section, we delve into the intricacies of each organism, examining their unique adaptations and the roles they play within the delicate balance of the colony's ecosystem. From the genetic modifications that enhance their resilience to the careful monitoring of their interactions, every aspect of these life forms is scrutinized to ensure the sustainability of the colony.

The essence of biodiversity in space is not just a scientific marvel; it's a testament to human innovation and our ability to adapt and thrive even in the most inhospitable environments. As we explore the intricate web of life within space colonies, we gain a deeper understanding of our connection to the cosmos and the endless possibilities that the future of space colonization holds.

Challenges and Innovations in Maintaining Biodiversity

Within the limited confines of space colonies, the task of preserving biodiversity transforms into a formidable challenge that demands unparalleled innovation. Imagine vertical farms stretching majestically toward the ceiling, their tiers adorned with a multitude of genetically modified crops meticulously designed to thrive in alien soils. These crops are not just agricultural marvels; they represent a testament to human ingenuity, a fusion of science and nature that ensures the continuity of life beyond Earth.

Aquatic ecosystems, carefully balanced and meticulously monitored, harbor a diverse array of organisms, each one meticulously engineered to coexist harmoniously within the controlled confines of space. These organisms, adapted through the wonders of genetic engineering, are the pioneers of extraterrestrial agriculture. They are more than simple inhabitants; they are the lifeblood of space colonies, providing sustenance, oxygen, and vital nutrients to the inhabitants.

In this part of our exploration, we venture into the heart of these innovative solutions. We witness the delicate dance between human creativity and the forces of nature, as scientists and engineers push the boundaries of what is possible. The vertical farms, illuminated by the soft glow of artificial light, are a testament to our ability to adapt agriculture to extraterrestrial conditions. Here, plants are not bound by seasons or natural sunlight; instead, they flourish under carefully calibrated LED lights, their growth optimized for space.

Genetic modification, once a contentious topic, has become a cornerstone of space agriculture. Scientists have meticulously crafted crops resistant to harsh environments, fortified with essential nutrients, and engineered to thrive in alien soils. These genetically modified organisms represent a new frontier in agriculture, offering solutions to challenges that were once insurmountable.

Aquatic ecosystems, too, have undergone a transformation. In carefully designed tanks, engineered microorganisms recycle nutrients, ensuring that no resource goes to waste. Fish, genetically modified for rapid growth and efficient nutrient conversion, swim alongside algae engineered to produce oxygen at unprecedented rates. These aquatic marvels are not only a source of sustenance but also a model of efficient resource utilization, showcasing the brilliance of nature enhanced by human intervention.

The marriage of human ingenuity and nature in space colonies is not just a scientific achievement;

it's a testament to our determination to thrive beyond the confines of Earth. As we explore these innovative solutions, we gain insights into the delicate balance between adaptation and innovation. The challenges of maintaining biodiversity in space are met with creativity and resilience, paving the way for a future where life not only survives but thrives in the vast expanse of the cosmos.

Genetically Modified Organisms: Guardians of Ecological Balance

At the core of preserving biodiversity in space lie the guardians of ecological balance: Genetically Modified Organisms (GMOs). These specially crafted life forms stand as marvels of genetic engineering, their DNA intricately adjusted to withstand the rigors of extraterrestrial habitats. In this section, we embark on a profound journey into the world of GMOs, unraveling the secrets of their creation, understanding their pivotal role in supporting ecosystems, and grappling with the ethical considerations that accompany their existence. Here, we confront the delicate balance between intervention and preservation, delving into the moral quandaries of playing nature's architect.

Genetic modification, once confined to the realms of science fiction, has become an indispensable tool for space colonization. Scientists, armed with a deep understanding of genetics and a profound respect for nature, have meticulously designed GMOs capable of thriving in environments previously deemed inhospitable. These organisms are not just products of laboratory manipulation; they are living solutions to the challenges of space.

In the creation of GMOs, precision is paramount. Scientists target specific genes, enhancing traits that enable survival in extraterrestrial conditions. Resistance to radiation, tolerance to extreme temperatures, and efficient nutrient utilization are just a few examples of the traits bestowed upon these modified organisms. Through careful manipulation, GMOs become resilient pioneers, adapting to the unique challenges posed by space habitats.

Yet, the creation of GMOs is not without ethical complexities. As we engineer organisms to fit our needs, questions arise about the consequences of our interventions. Are we tampering with the natural order, or are we acting as stewards of life's evolution? This section delves deeply into these ethical considerations, exploring the fine line between scientific advancement and environmental ethics. We confront the moral dilemmas of altering organisms to preserve biodiversity, weighing the benefits against the potential risks.

The role of GMOs in supporting ecosystems within space colonies is multifaceted. In controlled environments, these organisms contribute to the intricate web of life, serving as sources of food, oxygen, and vital nutrients. Algae engineered to produce oxygen, plants fortified with essential vitamins, and microorganisms recycling nutrients exemplify the diverse array of GMOs essential for extraterrestrial habitats.

As we navigate the realm of genetically modified organisms, we gain insights into the complex interplay between human intervention and ecological balance. We grapple with profound questions about our responsibility as architects of life, considering the legacy we leave for future generations. In this exploration, we confront the ethical challenges head-on, acknowledging the power of science to shape the destiny of life beyond Earth.

The Ethical Balance of Introducing Earth Species

In the vast expanse of space colonization, one of the most monumental decisions humanity faces is the introduction of Earth species to alien worlds. This pivotal act, essential for preserving biodiversity, carries with it profound ethical implications that resonate across the cosmos. As humanity steps into the role of cosmic gardener, nurturing life in extraterrestrial environments, a myriad of questions permeates our collective conscience. What rights do we possess to usher in Earth's biodiversity to distant planets? What responsibility do we bear for the consequences, both intended and unintended, of our actions? In this section, we plunge into the depths of the ethical abyss, challenging our established perceptions and urging contemplation. Here, we confront the moral dilemmas inherent in our newfound role as stewards of life in the cosmos.

The act of introducing Earth species to alien worlds signifies a transformative leap in our relationship with the universe. It marks a departure from being mere spectators of the cosmos to becoming active participants, shaping the very fabric of life beyond our home planet. This monumental decision hinges on the delicate balance between our desire to preserve biodiversity and the potential ecological impacts of our interventions.

At the heart of the ethical dilemma lies the question of interference. As custodians of life, do we have the right to meddle with the natural evolution of extraterrestrial ecosystems? By introducing Earth organisms, are we not disrupting the delicate equilibrium that has existed on these alien worlds for millennia? These ethical quandaries demand introspection and a nuanced understanding of our place in the grand tapestry of the universe.

Furthermore, the responsibility we bear for the consequences of our actions cannot be understated. Every introduction of an Earth species carries with it the potential for unintended ecological consequences. Predatory species could disrupt existing food chains, while fast-reproducing organisms might overrun native flora and fauna. The ethical tightrope we walk demands careful consideration of the potential ripple effects of our decisions.

In this exploration of ethical dimensions, we grapple with the very essence of our humanity. We question our ethical obligations not just to Earth but to the universe at large. As we venture into the unknown, we are tasked with upholding the principles of respect, humility, and responsibility. This section does not provide definitive answers but invites readers to engage in the profound ethical discourse that space colonization demands. Here, we confront the complexities of our actions, acknowledging the weight of our decisions as we navigate the uncharted territories of life in the cosmos.

In these pages, the complexities of biodiversity preservation unfold. Each word is a step into the unknown, a contemplation of the intricate web of life in space colonies. There are no easy answers, only a profound awareness of the ethical tightrope humanity walks as it ventures into the unexplored realms of space. The journey continues, inviting readers to grapple with the moral intricacies of preserving biodiversity beyond Earth.

Chapter 22: Space Colonies and Artistic Expression

In the boundless realms of space colonies, where technology and nature harmoniously coexist, a vibrant tapestry of artistic expression unfolds. Far from the confines of Earth, creativity takes flight, transcending the ordinary and reaching for the cosmic unknown. This chapter embarks on a captivating journey into the heart of artistic ingenuity within the extraterrestrial havens, exploring the myriad ways in which art and creativity flourish amidst the stars.

The Cosmic Canvas: Space-Inspired Art Forms

Within the sterile walls of space habitats, artists wield their brushes as instruments of cosmic exploration. Space-inspired art forms come to life, depicting the awe-inspiring grandeur of distant galaxies, the dance of celestial bodies, and the mystique of uncharted worlds. In this section, we delve into the world of space art, exploring the intricate techniques employed by artists to capture the ethereal beauty of the cosmos. Through intricate brushstrokes and vivid palettes, they bring the wonders of the universe closer to the hearts of colonists, fostering a profound sense of connection with the cosmos.

Within the confines of space habitats, where sterility meets the human spirit, artists embark on a celestial odyssey, their brushes tracing the mysteries of the universe. Space-inspired art forms, born in the silent chambers of colonies, transcend the boundaries of imagination, capturing the essence of cosmic exploration in strokes of brilliance.

Capturing the Awe-Inspiring Grandeur

In this cosmic gallery, artists harness the power of their creativity to depict the awe-inspiring grandeur of distant galaxies. With meticulous detail, they craft spiraling nebulae, painting the interstellar clouds with hues unseen by human eyes. Through their artistry, the vastness of space unfolds, inviting viewers to peer into the depths of the cosmos and wonder at the infinite possibilities that lie beyond.

The Dance of Celestial Bodies

With every stroke, artists choreograph the celestial ballet, where stars pirouette in cosmic rhythms. Moons and planets waltz around their parent stars, creating mesmerizing patterns in the cosmic dance floor. Artists capture this harmony, freezing moments of stellar elegance on canvases. Through their art, the dance of celestial bodies becomes a visual symphony, inspiring awe and reverence for the cosmic order that governs the universe.

The Mystique of Uncharted Worlds

In the uncharted realms of space, where the unknown beckons the curious, artists venture into the realms of speculative art. They imagine alien worlds, each unique in its landscapes, atmospheres, and inhabitants. These creations, born from the depths of artistic imagination, fuel the dreams of colonists and astronauts alike. Through these artworks, the mystique of uncharted worlds finds expression, igniting the spirit of exploration that defines humanity's cosmic journey.

Fostering Connection with the Cosmos

Space art transcends the visual realm; it fosters a profound connection between colonists and the cosmos. These creations adorn the walls of space habitats, reminding inhabitants of the boundless wonders that lie beyond their metallic confines. Artists infuse life into sterile environments, infusing them with the vibrancy of the universe. Through their art, colonists find solace in the vastness of space, forging an intimate bond with the stars, planets, and galaxies that adorn their artistic expressions.

Harmonies of the Void: Music Compositions in Space

In the silent expanse of space, music becomes a universal language, resonating through the corridors of space colonies. Musicians, inspired by the celestial ballet above, compose symphonies that echo the cosmic rhythms. This segment delves into the ethereal compositions born in the void, exploring the fusion of earthly melodies with the celestial harmonies of the universe. From ambient space music that mimics the hum of distant stars to orchestral pieces inspired by the majesty of planetary orbits, music becomes a bridge between humanity and the cosmos.

In the profound silence of the cosmic void, where sound dissipates into nothingness, music emerges as a universal language, echoing through the corridors of space colonies. Within these metal and glass confines, musicians find inspiration in the celestial ballet above, weaving symphonies that transcend earthly boundaries and resonate with the very essence of the universe.

The Universal Language of Music

In space colonies, music becomes a bridge between humanity and the cosmos. Musicians, attuned to the harmonies of the universe, compose ethereal pieces that capture the essence of the celestial void. These compositions traverse the vast distances of space, reaching the hearts of colonists and astronauts alike. From the subtle vibrations of ambient space music, mimicking the distant hum of stars, to grand orchestral pieces inspired by the majesty of planetary orbits, music transcends the confines of the colonies, allowing inhabitants to experience the cosmic dance in auditory form.

Echoes of Celestial Rhythms

Inspired by the rhythmic dance of celestial bodies, musicians craft compositions that echo the cosmic harmonies. The gentle ebb and flow of planets around their stars find resonance in the delicate melodies played on futuristic instruments. Percussive beats mimic the pulsars' rhythmic pulses, creating a symphony that mirrors the heartbeat of the universe. These musical creations, born in the silent expanse of space, pay homage to the grandeur of the cosmos, allowing listeners to immerse themselves in the profound harmony of the void.

Fusion of Earthly Melodies and Celestial Harmonies

In this cosmic symphony, earthly melodies find harmony with celestial tunes. Musicians experiment with instruments and sounds, blending terrestrial rhythms with the harmonies of the cosmos. The result is a fusion of the familiar and the extraordinary, where traditional notes intermingle with the ethereal hum of interstellar space. Through this fusion, music in space colonies becomes a testament to the interconnectedness of the universe, reminding inhabitants of their place in the cosmic order.

The Ethereal Compositions Born in the Void

Within the void, where sound takes on a new dimension, musicians explore uncharted territories of composition. They create pieces that transcend conventional musical boundaries, embracing the ethereal and the surreal. From ambient tunes that capture the tranquility of interstellar space to avant-garde compositions that challenge the very definition of music, these ethereal pieces become sonic journeys, guiding listeners through the vast expanses of the cosmos.

In this segment, we delve into the ethereal compositions born in the void, exploring the fusion of earthly melodies with the celestial harmonies of the universe. From ambient space music that mimics the hum of distant stars to orchestral pieces inspired by the majesty of planetary orbits, music becomes a bridge between humanity and the cosmos, resonating with the universal language that unites all beings in the cosmic tapestry.

Words Among the Stars: Literary Works in Extraterrestrial Solitude

In the solitude of space colonies, pens dance across pages, weaving narratives that explore the complexities of life beyond Earth. Science fiction, once a genre rooted in Earthly imagination, now gains new dimensions in the cosmic expanse. This part of our exploration immerses readers in literary works crafted by colonists, delving into futuristic tales of interstellar voyages, alien encounters, and the intricacies of human nature in the vastness of space. From epic space operas to introspective poetry inspired by the isolation of space, literature becomes a means for colonists to grapple with the profound questions of existence.

In the vast solitude of space colonies, where silence stretches endlessly, pens dance across pages, giving birth to narratives that explore the profound mysteries of life beyond Earth. Science fiction, once confined to the realms of Earthly imagination, gains new dimensions in the cosmic expanse. In this segment of our exploration, we invite readers to immerse themselves in literary works crafted by colonists, delving into the intricacies of interstellar voyages, encountering extraterrestrial life, and exploring the depths of human nature against the backdrop of the cosmic void.

The Cosmic Chronicles: Exploring Interstellar Voyages

Within the quiet confines of space colonies, authors craft epic space operas that chronicle the adventures of intrepid explorers venturing into the unknown cosmos. These narratives take readers on interstellar voyages, where spacecraft traverse galaxies, encountering alien civilizations and navigating cosmic anomalies. Through the vivid imagination of writers, readers are propelled into the heart of unexplored worlds, exploring the wonders and perils of the universe, all from the safety of their space colony dwellings.

Encounters Beyond Earth: Exploring Alien Contact

In the literary tapestry of space colonies, tales of alien encounters abound. Colonist-authors weave intricate stories of humanity's first contact with extraterrestrial beings. These narratives delve into the complexities of communication, the clash of cultures, and the profound questions that arise when humanity encounters life beyond Earth. Through these imaginative explorations, readers confront the unknown, pondering the potential similarities and vast differences between terrestrial life and alien civilizations.

Introspective Echoes: Poetry in the Cosmic Silence

Amidst the isolation of space, poets find solace in crafting verses that echo the solitude and grandeur of the cosmos. Their introspective poetry delves into the isolation of space, the vastness of the universe, and the existential questions that arise in the silence of the void. Through poetic language, these colonist-poets explore the human condition in the face of cosmic infinitude, capturing the emotions and contemplations that arise when humanity confronts the boundless expanse of space.

Human Nature in the Vastness of Space: Exploring Philosophical Themes

In the literary works of space colonies, philosophical themes intertwine with the cosmic narratives. Authors delve into the depths of human nature, exploring existential questions, morality, and the search for meaning in the cosmic tapestry. Through these explorations, readers are prompted to reflect on their own existence, contemplating the profound philosophical inquiries that arise when humanity ventures into the unknown reaches of space.

In this segment, we have delved into literary works in the solitude of space colonies, exploring interstellar voyages, encounters with extraterrestrial life, introspective poetry, and philosophical themes. Through the written word, colonists grapple with the complexities of existence, inviting readers to join them on a literary journey that spans the vast cosmic expanse.

The Fusion of Science and Art: Enhancing Life in Space

Amidst the sterile pragmatism of life in space, art becomes a beacon of emotional resonance and cultural identity. This section illuminates the vital role of artistic expression in enhancing the quality of life within space colonies. From aesthetically designed habitats that evoke a sense of harmony to immersive virtual reality experiences that blend art and technology, creativity infuses every aspect of colony life. The fusion of science and art not only enriches the sensory experiences of colonists but also fosters a deep appreciation for the beauty of the universe they inhabit.

In the sterile pragmatism of life within space colonies, art emerges as a profound beacon of emotional resonance and cultural identity. In this segment, we delve into the symbiotic relationship between science and art, exploring how artistic expression becomes an integral part of enhancing the quality of life for colonists.

Aesthetically Designed Habitats: Harmonizing Space and Artistry

Within the confines of space habitats, architects and artists collaborate to create aesthetically designed living spaces. These habitats are more than mere functional structures; they are artistic marvels that evoke a sense of harmony and beauty. Walls adorned with murals depicting cosmic vistas, ceilings mimicking the dance of stars, and common areas designed as interactive art installations transform the sterile environment into a canvas of creativity. Through these carefully crafted spaces, colonists find solace and inspiration, fostering a sense of connection with the vast universe beyond their colony walls.

Immersive Virtual Reality Experiences: Blending Art and Technology

In the realm of virtual reality, art and technology merge seamlessly, offering colonists immersive experiences that transcend the boundaries of physical space. Virtual reality simulations transport individuals to fantastical worlds, allowing them to explore artistic creations that defy the constraints of reality. Artists collaborate with software developers to craft virtual landscapes inspired by cosmic wonders, enabling colonists to embark on virtual journeys through galaxies, nebulae, and alien worlds. Through these immersive encounters, colonists not only escape the confines of their immediate surroundings but also gain a deeper appreciation for the artistic beauty inherent in the universe.

Artistic Exploration of Microgravity: Capturing Weightless Beauty

In the unique environment of microgravity, artists explore new dimensions of creativity. Floating effortlessly within their colony modules, artists experiment with techniques that are impossible in Earth's gravity. Floating sculptures, kinetic art pieces, and interactive installations come to life, capturing the weightless beauty of space. These creations serve not only as expressions of artistic vision but also as reminders of the extraordinary environment in which colonists reside. Through the artistic exploration of microgravity, colonists find inspiration in the novel experiences offered by their cosmic abode.

The Emotional Resonance of Art: Nurturing the Human Spirit

Beyond aesthetics, art within space colonies holds a profound emotional resonance. Creative expressions, whether visual art, music, or literature, become a source of comfort and solace in the face of isolation. Artistic performances and exhibitions bring colonists together, fostering a sense of community and shared cultural identity. Through these shared experiences, the human spirit is nurtured, and a deep appreciation for the beauty of the universe permeates the collective consciousness of the colony inhabitants.

In this segment, we have explored the fusion of science and art within space colonies, delving into aesthetically designed habitats, immersive virtual reality experiences, artistic exploration of microgravity, and the emotional resonance of art. Through this integration of creativity and scientific ingenuity, life in space colonies becomes not only functional but also profoundly enriching, enhancing the sensory and cultural experiences of the colonists as they navigate the cosmic frontier.

As we journey through this chapter, we witness the flourishing of artistic expression in the extraterrestrial abodes. Through the strokes of brushes, the melodies of instruments, and the power of words, colonists find solace, inspiration, and a profound connection with the cosmos. Art becomes a testament to the enduring spirit of humanity, transcending the boundaries of Earth and reaching for the stars. In the cosmic canvases of space colonies, creativity knows no bounds, painting the universe with the vibrant hues of human imagination.

Chapter 23: Space Colonies and Quantum Computing

Quantum Computing Fundamentals

Quantum computing, a realm governed by the laws of quantum mechanics, introduces a paradigm shift in computational science. In the context of space colonies, this technology offers groundbreaking solutions to complex challenges. Quantum bits, or qubits, exist in states of superposition, enabling the simultaneous processing of multiple possibilities. Entanglement, a phenomenon where qubits become interconnected, further enhances computational efficiency. These principles allow quantum computers to explore vast solution spaces at speeds unattainable by classical computers.

Quantum Computing Architectures

Space colonies demand computing systems resilient to cosmic rigors. Quantum computing architectures, each with unique attributes, stand as pillars of this computational revolution. Superconducting qubits, utilizing superconducting circuits, offer stability and scalability. Trapped ions, manipulated using electromagnetic fields, ensure high precision. Topological qubits, resilient against errors, promise fault-tolerant computations. Quantum dots, utilizing semiconductor technology, provide compact solutions. Adapting these architectures to space conditions poses challenges, but innovators strive to conquer these obstacles.

Quantum Computing Applications in Space Simulations

In the realm of space simulations, quantum computing reigns supreme. Traditional simulations, essential for understanding celestial phenomena, demand enormous computational resources. Quantum algorithms, designed for parallel processing, accelerate these simulations manifold. Space colonies utilize quantum computers to model intricate gravitational interactions, simulate planetary formations, and predict space weather patterns. These applications not only enhance scientific understanding but also inform strategic decisions crucial for space missions.

Quantum Cryptography and Secure Communication

Securing communication channels in space is paramount. Quantum cryptography exploits the principles of quantum mechanics to create inherently secure communication networks. Quantum key distribution (QKD) protocols, based on the unique properties of quantum states, enable the exchange of cryptographic keys securely.

Space colonies leverage QKD to establish encrypted communication links, safeguarding sensitive data transmitted between colonies and Earth. The unbreakable nature of quantum encryption ensures the confidentiality and integrity of interplanetary communications.

Quantum Machine Learning for Data Analysis

Data analysis forms the backbone of decision-making processes within space colonies. Quantum machine learning algorithms, adept at processing vast datasets, offer unparalleled insights. Quantum algorithms, such as the Quantum Support Vector Machine (QSVM) and Quantum Principal Component Analysis (QPCA), outperform classical counterparts in analyzing complex data patterns. Space colonists employ quantum machine learning to optimize resource allocation, predict equipment failures, and analyze biological experiments. The fusion of quantum computing and machine learning elevates the colony's analytical capabilities to unprecedented heights.

Quantum Computing in Materials Science and Manufacturing

Materials science advancements are pivotal for sustainable space colonization. Quantum computing accelerates materials discovery and optimization. Quantum simulations elucidate the behavior of materials under extreme conditions, aiding in the development of spacecraft components resistant to cosmic radiation and temperature fluctuations. Moreover, quantum algorithms optimize manufacturing processes, reducing waste and energy consumption. Space colonies harness these capabilities to create advanced materials, ensuring the longevity and resilience of their habitats.

Quantum Computing and Artificial Intelligence Integration

The integration of quantum computing and artificial intelligence heralds a new era of intelligent decision-making. Quantum neural networks and quantum reinforcement learning algorithms optimize autonomous systems. Quantum computing accelerates AI training processes, enabling rapid adaptation to dynamic environments. In space colonies, AI-driven systems equipped with quantum processing capabilities manage habitat environments, control robotics, and assist in scientific research. This synergy enhances the colony's efficiency, enabling adaptive responses to unforeseen challenges.

Quantum Computing Ethics and Governance

The advent of quantum computing raises ethical questions and governance challenges. Quantum computing's potential to break existing cryptographic systems poses security risks if misused. Space colonies establish robust ethical frameworks, ensuring responsible quantum technology use. International cooperation fosters quantum technology governance standards, preventing misuse and ensuring a peaceful coexistence of quantum-powered civilizations in the cosmos.

In the cosmic tapestry of space colonization, quantum computing emerges as a guiding star, illuminating pathways to scientific discovery, secure communication, advanced materials, and artificial intelligence integration. As space colonies continue their celestial odyssey, the fusion of quantum computing and human ingenuity propels them toward unparalleled frontiers of knowledge and exploration.

Chapter 24: Space Colonies and Sustainable Energy

In the boundless expanse of space, energy stands as the lifeblood of thriving colonies. Sustainable energy solutions, paramount for the longevity and self-sufficiency of space habitats, pave the way toward a harmonious coexistence with the cosmos. Within the confines of space colonies, innovative approaches to harnessing energy emerge, shaping the future of extraterrestrial civilizations.

Solar Power: The Radiant Lifeline

Solar power stands as a beacon of renewable energy in space colonies. Sunlight, abundant in the cosmic void, becomes the primary source of power. Advanced solar panels, leveraging photovoltaic cells and concentrated solar power techniques, capture solar energy with unprecedented efficiency. Space-based solar arrays, orbiting celestial bodies, provide uninterrupted power by avoiding planetary eclipses. Space colonies unfurl vast arrays of glistening solar panels, transforming sunlight into the energy that illuminates their habitats and powers their scientific endeavors.

Nuclear Energy: The Cosmic Furnace

Nuclear energy, harnessed through compact nuclear reactors, offers a potent energy source for space colonies. Radioisotope thermoelectric generators (RTGs), fueled by decaying isotopes, provide steady power for deep space missions and extraterrestrial research stations. Advanced fission reactors, employing innovative cooling systems, supply colonies with abundant energy. Space colonies, equipped with nuclear reactors, navigate the cosmos, their reactors glowing softly with the promise of sustainable power generation.

Innovative Technologies: Space-Based Solar Arrays and Beyond

In the quest for energy sustainability, space colonies pioneer groundbreaking technologies. Space-based solar arrays, composed of interconnected modules, capture sunlight beyond Earth's atmosphere, eliminating atmospheric interference. Advanced rectennas, converting microwave energy transmitted from solar power satellites, wirelessly transfer energy to colonies' power grids. Fusion reactors, emulating the sun's energy production, promise limitless clean energy. Space colonies venture into the realm of fusion, aiming to unlock the potential of this revolutionary energy source.

Energy Storage and Distribution Challenges

Energy storage and distribution pose intricate challenges in space environments. Advanced energy storage systems, utilizing high-capacity batteries and supercapacitors, store excess energy during peak production periods. Superconducting transmission lines minimize energy loss during distribution. Space colonies meticulously design their energy grids, balancing supply and demand while ensuring resilience against unforeseen fluctuations. Redundant energy storage and distribution networks guarantee uninterrupted power supply, essential for sustaining life and scientific pursuits.

Harnessing Planetary Resources: Mining the Energy Wealth of Celestial Bodies

Planetary resources, abundant in the form of minerals and volatiles, hold the key to sustainable energy production. Space colonies establish mining operations on moons, asteroids, and planetary surfaces, extracting essential elements for energy generation. Helium-3, a rare isotope abundant on the moon, fuels fusion reactors. Water ice, prevalent in asteroids, undergoes electrolysis to produce hydrogen for fusion and oxygen for life support. Space colonies, miners of celestial bodies, tap into the energy wealth of the cosmos, ensuring their autonomy and long-term energy sustainability.

In the cosmic tapestry of space colonization, sustainable energy solutions illuminate the path toward self-reliance and environmental stewardship. Solar power, nuclear energy, innovative technologies, and the harnessing of planetary resources converge to energize the ambitions of spacefaring civilizations. As space colonies thrive amidst the stars, their commitment to sustainable energy illuminates the darkness of the cosmic void, symbolizing humanity's ability to harmonize with the universe while advancing the frontiers of knowledge and exploration.

Chapter 25: Space Colonies and Interstellar Communication

In the vast cosmic theater, communication stands as the thread that weaves together the tapestry of interstellar civilizations. For space colonies, the challenges of transmitting information across astronomical distances pose intricate puzzles, pushing the boundaries of human ingenuity and technology. This chapter delves into the complexities of interstellar communication, unraveling the limitations, innovative solutions, and societal ramifications faced by space colonies as they reach out to the stars.

The Limits of Light-Speed Communication: Bridging Cosmic Distances

Light-speed communication, the cosmic courier of information, encounters formidable limitations in the vastness of space. As space colonies venture farther into the cosmos, the delay in communication becomes palpable. Messages sent from distant colonies take years, even centuries, to reach Earth. The constraints of light-speed communication create a chasm of isolation, prompting spacefaring civilizations to seek novel methods to bridge the temporal gaps between transmissions.

Quantum Entanglement: The Enigmatic Messenger

Quantum entanglement, a phenomenon that defies classical physics, emerges as a beacon of hope in the realm of interstellar communication. Entangled particles, connected regardless of distance, offer the promise of instantaneous communication. Space colonies delve into the intricacies of quantum entanglement, harnessing its mysterious properties to transmit information faster than the speed of light. Quantum communication devices, employing entangled particles as information carriers, become the cornerstone of interstellar communication networks, transcending the limitations of traditional methods.

Communication Protocols: Navigating the Celestial Silence

Interstellar communication necessitates meticulous protocols and strategies. Colonies develop robust encryption methods to secure transmissions, protecting sensitive information from interception. Error correction algorithms, designed to counter data degradation during vast cosmic journeys, ensure the integrity of transmitted messages. Space colonies establish a standardized interstellar communication language, enabling seamless exchanges between Earth and distant civilizations. The development of autonomous communication systems allows colonies to adapt and respond to unforeseen challenges, fostering resilience in the face of cosmic uncertainties.

Cultural and Societal Impacts: The Echoes of Silence

Delayed communication leaves an indelible mark on the cultural and societal fabric of space colonies. The temporal disconnect between colonies and Earth gives rise to unique cultural identities, shaped by isolation and self-reliance. Colonists, distanced from terrestrial events, form their traditions, arts, and social norms, fostering a sense of unity within their cosmic communities. The prolonged communication delays evoke introspection and contemplation, prompting profound philosophical and existential questions among spacefaring civilizations. The interplay between isolation and connectedness molds the collective psyche of interstellar communities, painting a nuanced portrait of humanity's resilience in the face of cosmic solitude.

In the cosmic ballet of interstellar communication, space colonies navigate the vastness of space and time, seeking to bridge the gaps between stars. Light-speed communication limitations, quantum entanglement mysteries, meticulous communication protocols, and the cultural impacts of delayed transmissions form the backdrop of this chapter. As space colonies strive to communicate across the cosmic expanse, they unravel the enigmas of the universe, embracing the challenges and opportunities that interstellar communication presents on their journey toward cosmic understanding and unity.

Chapter 26: Space Colonies and Cultural Exchange with Earth

In the cosmic dance of cultural exchange, space colonies emerge as vibrant hubs of diversity and innovation, fostering connections with Earth that bridge the vastness of space. This chapter delves into the intricate tapestry of cultural exchange programs between space colonies and Earth, exploring the exchange of knowledge, traditions, and art forms that transcend the cosmic void. Amidst the challenges of millions of miles of separation, spacefaring civilizations navigate the complexities of maintaining a profound connection with Earth's ever-evolving cultural landscape.

The Interstellar Dialogue: Exchanging Knowledge Across Cosmic Frontiers

Cultural exchange programs become portals of interstellar dialogue, enabling the sharing of knowledge between Earth and space colonies. Scientists, scholars, and artists from Earth collaborate with spacefaring civilizations, engaging in virtual lectures, workshops, and collaborative research projects. Virtual classrooms stretch across astronomical distances, bringing together students from Earth and space colonies, fostering a global community of learners united by their passion for knowledge. Space colonies become centers of wisdom, where Earth's intellectual heritage merges with cosmic insights, shaping the minds of future generations.

Traditions Across Light-Years: Preserving Earth's Heritage in Space

Cultural exchange extends beyond intellectual pursuits, encompassing the rich tapestry of traditions from Earth's diverse cultures. Space colonies celebrate Earth's festivals, honoring traditions that echo through millennia. From Diwali's luminous celebrations to the vibrant rhythms of Carnaval, spacefaring civilizations embrace Earth's cultural heritage, weaving these traditions into the fabric of their cosmic communities. Colonists engage in traditional arts and crafts, culinary experiences, and storytelling sessions, preserving Earth's cultural legacy amidst the stars. These traditions not only connect space colonies with Earth but also become cherished aspects of interstellar life, fostering a sense of shared humanity.

Artistic Alchemy: Cosmic Creations Inspired by Earth's Cultures

Art becomes a universal language in the interstellar cultural exchange. Artists from Earth and space colonies collaborate, drawing inspiration from Earth's diverse cultures to create masterpieces that transcend terrestrial boundaries. Music compositions blend the melodies of Earth's cultures with cosmic rhythms, creating harmonies that resonate across the cosmic expanse. Visual arts, from paintings to sculptures, reflect the kaleidoscope of Earth's artistic traditions, adorning the walls of space colonies with vibrant colors and profound symbolism. Literature and poetry intertwine, weaving narratives that explore the human experience in the cosmic context. The artistic alchemy between Earth and space colonies births creations that inspire, provoke, and evoke a deep sense of connection with the universe.

Cultural Challenges: Navigating Earth's Evolving Landscape

Maintaining a connection with Earth's evolving cultural landscape presents unique challenges. Space colonies grapple with the rapid pace of cultural evolution on Earth, adapting their cultural exchange programs to embrace emerging trends and movements. Virtual reality technologies create immersive experiences, allowing space colonists to virtually visit Earth's cultural events, museums, and heritage sites. Language evolves, necessitating continuous linguistic adaptations to foster effective communication. Space colonies become cultural chameleons, adept at navigating Earth's diverse cultural terrains, ensuring that the interstellar dialogue remains vibrant and relevant.

The Human Connection: Nurturing Bonds Across Cosmic Chasms

At the heart of cultural exchange lies the profound human connection between Earth and space colonies. Individuals from diverse backgrounds, separated by millions of miles, find common ground in shared experiences, emotions, and aspirations. The exchange of personal stories, family traditions, and everyday life creates a deep sense of empathy and understanding. Through virtual reality gatherings, video calls, and collaborative projects, individuals on Earth and in space colonies forge bonds that transcend the celestial distances. The human connection becomes the foundation upon which interstellar friendships and collaborations flourish, enriching the lives of both Earthlings and spacefarers.

In the boundless expanse of space, cultural exchange programs become beacons of connection, weaving the threads of knowledge, traditions, and artistry between Earth and space colonies. The interstellar dialogue, the preservation of traditions, the artistic alchemy, the cultural challenges, and the human connection form the intricate mosaic of this chapter, depicting the resilience, creativity, and shared humanity that define the cultural exchange between Earth and the cosmic frontier.

Chapter 27: Space Colonies and Law Enforcement

In the vast expanse of space, law enforcement takes on new dimensions, addressing the unique challenges of maintaining order and security within isolated space colonies.

Guardians of Cosmic Peace: Space Police and Their Role

Space Police, the dedicated law enforcement agencies of space colonies, are the custodians of cosmic peace. Trained in zero-gravity combat and equipped with advanced technologies, they ensure the safety and well-being of colonists. Their duties encompass crime prevention, conflict resolution, and emergency response. These officers are not just enforcers of law; they are community builders, fostering a sense of unity among colonists and establishing protocols that promote harmony.

Crime Prevention Strategies: A Community Approach

Crime prevention in space colonies involves a community-focused approach. Colonists engage in conflict resolution workshops, where communication and empathy are emphasized. Psychological support services are readily available, addressing the challenges of social isolation and cabin fever. Surveillance technologies play a vital role, not just in identifying potential threats, but also in deterring criminal activities. AI-driven predictive algorithms analyze behavioral patterns, helping Space Police intervene proactively. Community engagement and education serve as the foundation, creating a cohesive environment where colonists actively contribute to their safety.

Conflict Resolution in Microgravity: Unique Challenges and Solutions

In the absence of gravity, conflict resolution takes innovative forms. Mediators undergo specialized training, incorporating techniques like zero-gravity yoga and mindfulness practices. Neutral spaces equipped with holographic interfaces facilitate conflict resolution sessions. These environments promote a sense of neutrality, enabling parties to express themselves openly. Mediation sessions often incorporate virtual reality simulations, allowing colonists to visualize their concerns and find common ground. Restorative justice programs focus on healing rather than punitive measures, fostering a sense of accountability and understanding among offenders.

Advanced Surveillance Technologies: Safeguarding Colonists

Space colonies rely on cutting-edge surveillance technologies for security. AI-driven surveillance cameras continuously monitor activities, analyzing data in real-time. Biometric identification systems, including facial recognition and retinal scans, ensure secure access control. Smart sensors detect anomalies in environmental parameters, alerting authorities to potential hazards. The ethical use of these technologies is paramount. Personal data is anonymized, and stringent regulations govern their deployment. Transparency and open communication between law enforcement agencies and colonists build trust, ensuring that surveillance measures respect individual privacy.

Legal Justice in Space: Balancing Accountability and Rehabilitation

Space colonies prioritize rehabilitation and restorative justice. Offenders engage in educational programs, vocational training, and community service. Custodial sentences are rare, reserved for severe offenses. The focus is on reintegrating offenders into the community. Legal systems emphasize community healing, with victims and offenders participating in mediation processes. Colonies implement AI-driven decision support systems to assist judges in determining appropriate sentences, ensuring fairness and consistency. The emphasis is on creating an environment where colonists learn from their mistakes and contribute positively to society.

Emergency Response and Crisis Management: Preparedness in Space

Space colonies meticulously plan for emergencies, conducting regular simulation drills. Emergency response teams, equipped with AI-driven decision support systems, coordinate actions during crises. Evacuation protocols are in place, ensuring the safe relocation of colonists in emergencies. Crisis management involves collaboration between various departments, including medical, engineering, and law enforcement. Space Police play a crucial role in maintaining order during emergencies, ensuring a smooth evacuation process. The adaptive nature of emergency protocols allows colonies to respond effectively to unforeseen events, promoting resilience and safety.

International Collaboration in Law Enforcement: Navigating Jurisdictional Challenges

Law enforcement in space requires international cooperation. Colonies collaborate through mutual assistance treaties, sharing information and expertise. Jurisdictional challenges arise due to the international nature of space colonies.

International agreements define the scope of law enforcement jurisdiction, ensuring clarity in legal matters. Space Police undergo cross-cultural training, enhancing their ability to work seamlessly with diverse communities. Intercolonial collaboration involves joint patrols and information exchange, fostering a sense of unity among spacefaring civilizations. Legal experts from various colonies collaborate to address emerging challenges, ensuring a unified approach to law enforcement in space.

Promoting Ethical Policing: Ensuring Justice and Individual Rights

Ethical policing is paramount in space colonies. Transparency, accountability, and respect for individual rights are central tenets. Law enforcement agencies are governed by strict codes of conduct, with independent oversight bodies ensuring adherence to ethical standards. Colonists actively participate in community policing initiatives, promoting a sense of ownership and accountability. Officers undergo continuous training, emphasizing cultural sensitivity and unbiased policing. The promotion of justice and individual rights creates a secure environment where colonists thrive, confident in the integrity of their law enforcement agencies.

This chapter delves into the intricate workings of law enforcement in space colonies, exploring the multifaceted approaches that ensure safety, promote harmony, and uphold justice in the cosmic frontier.

Chapter 28: Space Colonies and Environmental Sustainability

In the pristine expanses of space colonies, environmental sustainability is not just a goal; it's a way of life. This chapter delves into the innovative strategies, cutting-edge technologies, and holistic approaches employed by spacefaring civilizations to preserve the delicate balance of ecosystems, conserve resources, and ensure a harmonious coexistence with the cosmos.

Creating Closed-Loop Ecosystems: Nature's Blueprint for Sustainability

Space colonies emulate nature's brilliance by creating closed-loop ecosystems. This section explores the intricacies of closed-loop systems, where waste products are recycled and repurposed, mirroring Earth's natural cycles. Colonies employ advanced technologies for water filtration, waste decomposition, and nutrient recycling. Aquaponic and hydroponic farming techniques maximize agricultural output while minimizing resource consumption. The closed-loop approach not only conserves resources but also fosters a deep connection between colonists and their environment, promoting responsible consumption and ecological awareness.

Harnessing Solar Energy: The Power of Cosmic Rays

Solar energy becomes the lifeblood of space colonies, powering habitats, sustenance systems, and scientific endeavors. This section delves into the utilization of space-based solar arrays, capturing cosmic rays and converting them into clean, renewable energy. Colonies strategically position solar panels on the exteriors of habitats, maximizing exposure to sunlight. Advanced energy storage technologies, such as graphene-based batteries, ensure a continuous power supply, even during periods of solar eclipse. The harnessing of solar energy not only reduces reliance on Earth-bound resources but also establishes a sustainable energy infrastructure for long-term space colonization.

Revolutionizing Waste Management: From Trash to Resources

Waste management in space colonies undergoes a revolutionary transformation. This section explores the innovative methods employed to convert waste into valuable resources. Colonies utilize bioreactors and microbial systems to break down organic waste, generating biogas for energy and nutrient-rich byproducts for agriculture.

Inorganic waste, including plastics and metals, undergo advanced recycling processes, transforming them into raw materials for 3D printing and construction. The zero-waste approach minimizes the ecological footprint of space colonies, ensuring that every resource is utilized to its fullest potential.

Space Farming: Cultivating Life Beyond Earth

Space farming becomes a cornerstone of sustainability, providing colonists with fresh produce and fostering a deep connection to nature. This section delves into the techniques of space farming, including hydroponics, aeroponics, and vertical farming. Colonies employ LED-based artificial lighting, tailored to specific plant growth stages, optimizing photosynthesis and crop yield. Genetically modified crops, engineered for space environments, thrive in controlled agricultural modules. Colonists actively participate in farming activities, cultivating a sense of responsibility and self-sufficiency. Space farming not only provides nutritious food but also contributes to psychological well-being, enhancing the overall quality of life in space colonies.

Exploring Planetary Resources: Mining the Cosmic Frontier

Space colonies venture into the exploration of planetary resources, tapping into the abundant wealth of celestial bodies. This section delves into asteroid mining, lunar excavation, and the extraction of rare minerals from planetary surfaces. Advanced robotics and autonomous mining equipment are deployed to gather resources, minimizing human exposure to hazardous environments. Colonies establish resource extraction bases on moons and asteroids, employing AI-driven systems to assess resource viability and optimize mining operations. The utilization of extraterrestrial resources reduces the strain on Earth's finite reserves, ensuring the long-term sustainability of space colonies.

Innovations in Water Recycling: The Elixir of Space Life

Water, the elixir of life, takes center stage in space colonies' sustainability efforts. This section explores advanced water recycling technologies, including electrocoagulation, membrane filtration, and vapor compression distillation. Colonies implement closed-loop water systems, where wastewater undergoes rigorous purification processes, transforming it into potable water. Smart sensors monitor water quality in real-time, ensuring its safety for consumption and other essential purposes. Water recycling efficiency reaches unprecedented levels, with colonies achieving near-perfect reuse rates. The conservation and responsible management of water resources become fundamental practices in space colonies, ensuring a continuous supply for colonists' needs.

Bioregenerative Life Support Systems: Nature-Inspired Solutions

Bioregenerative life support systems emulate nature's ingenious designs, creating self-sustaining ecosystems within colonies. This section explores the integration of plants, algae, and microbial communities into life support systems. Plants act as biological filters, absorbing carbon dioxide and releasing oxygen, while microorganisms break down organic waste and produce essential nutrients. Algae cultivation tanks provide supplemental oxygen and serve as a nutritious food source. Colonies employ advanced bioreactors, simulating natural ecosystems and ensuring a balanced exchange of gases and nutrients. Bioregenerative life support systems not only conserve resources but also enhance the psychological well-being of colonists, connecting them to the intricate web of life beyond Earth.

Sustainable Habitat Design: Balancing Comfort and Conservation

Space colonies revolutionize habitat design, striking a delicate balance between comfort and conservation. This section delves into innovative architectural approaches, utilizing lightweight materials and modular construction techniques.

Habitats incorporate green spaces, vertical gardens, and communal areas, fostering a sense of natural harmony within artificial environments. Smart climate control systems optimize energy usage, adjusting temperature and lighting based on colonists' needs and environmental conditions. Sustainable materials, including biodegradable plastics and recycled composites, are utilized in habitat construction, minimizing ecological impact. Colonists actively participate in habitat design, tailoring living spaces to accommodate individual preferences while promoting sustainability. The fusion of cutting-edge technology and ecological principles results in habitats that enhance quality of life, promote resource conservation, and integrate seamlessly with the surrounding environment.

Educational Initiatives: Fostering Environmental Awareness

Space colonies prioritize environmental education, fostering awareness and ecological consciousness among colonists. This section explores educational initiatives, including workshops, lectures, and hands-on activities focused on sustainability. Colonists engage in ecological research projects, studying the unique ecosystems of space habitats and contributing to scientific understanding. Educational modules incorporate ecological principles into various disciplines, nurturing a generation of scientists, engineers, and leaders committed to environmental preservation. Colonists actively participate in conservation efforts, conducting biodiversity surveys and implementing habitat restoration projects. Environmental education becomes a catalyst for cultural change, instilling a deep appreciation for nature and sustainability, shaping the values of spacefaring civilizations.

Promoting Interstellar Environmental Ethics: Stewardship of Cosmic Habitats

Space colonies extend their environmental ethics beyond their habitats, embracing the stewardship of cosmic habitats. This section delves into interstellar environmental ethics, emphasizing the responsible exploration and colonization of celestial bodies. Colonies adhere to strict protocols, avoiding contamination of extraterrestrial environments and preserving their pristine nature. International agreements define the ethical guidelines for planetary exploration, ensuring the protection of potential life forms and maintaining the scientific integrity of celestial bodies. Colonists participate in astrobiology research, studying extremophiles and microbial life in space habitats, enhancing our understanding of extraterrestrial ecosystems. Interstellar environmental ethics become the foundation of ethical space exploration, guiding colonists toward responsible and sustainable interactions with the cosmos.

This chapter illuminates the extraordinary efforts of space colonies in ensuring environmental sustainability, unveiling a tapestry of innovation, ecological consciousness, and responsible stewardship that defines the ecological legacy of humanity in the cosmos.

Chapter 29: Space Colonies and Human Augmentation

In the boundless realms of space, humanity ventures beyond the constraints of biology, embracing the frontiers of human augmentation. This chapter delves into the integration of cutting-edge human augmentation technologies within space colonies, exploring the enhancements designed to amplify astronaut capabilities, expand sensory experiences, and adapt humans to the challenges of cosmic environments.

Neural Interfaces: Bridging Minds and Machines

Neural interfaces emerge as revolutionary tools in human augmentation, bridging the gap between minds and machines. This section explores brain-computer interfaces (BCIs) and neuroprosthetics, enabling direct communication between the brain and external devices. Neural interfaces empower astronauts with seamless control over spacecraft systems, robotic companions, and scientific instruments. Colonists undergo neural augmentation procedures, enhancing cognitive abilities and enabling rapid information processing. Ethical considerations surround the integration of neural interfaces, raising questions about privacy, consent, and the potential for cognitive enhancement beyond natural human limits.

Augmented Senses: Expanding Perceptions in Space

Space colonization extends the boundaries of human perception through augmented senses. This section delves into enhancements such as bionic eyes, augmented hearing devices, and tactile sensory augmentation. Augmented senses grant astronauts unparalleled awareness of their surroundings, vital in the vastness of space. Visual enhancements enable real-time analysis of celestial phenomena, enhancing scientific observations. Auditory augmentation filters and amplifies sounds, allowing astronauts to perceive subtle cues in the spacecraft's machinery. Tactile augmentation provides haptic feedback, enabling precise manipulation of objects in microgravity. Ethical dilemmas arise concerning the enhancement of sensory experiences, sparking discussions about authenticity, altered perceptions, and the preservation of human identity.

Enhanced Physical Capabilities: Beyond Human Limits

Human augmentation extends beyond cognition, enhancing physical capabilities to adapt to the challenges of space environments. This section explores exoskeletons, muscle augmentation technologies, and artificial limbs designed for space. Exoskeletons support astronauts during extravehicular activities (EVAs), providing enhanced mobility and strength in microgravity. Muscle augmentation technologies prevent muscle atrophy, enabling prolonged space missions. Advanced prosthetic limbs restore mobility and dexterity, ensuring that individuals with limb loss can participate fully in space colony activities. Ethical considerations encompass issues of fairness, accessibility, and societal attitudes toward augmented individuals, reflecting on the inclusivity of human augmentation technologies.

Genetic Augmentation: Adapting the Human Blueprint

Genetic augmentation stands at the crossroads of ethical and scientific frontiers. This section navigates the ethical quandaries surrounding genetic modifications designed to enhance human adaptability in space. Genetic alterations enable resistance to cosmic radiation, immunity to diseases, and tolerance to extreme environmental conditions. Colonists undergo genetic therapies, ensuring the hereditary transmission of beneficial traits to future generations. Ethical debates center on the definition of humanity, consent for genetic modifications, and the potential consequences of altering the human genome. Society grapples with questions of diversity, identity, and the preservation of natural evolution in the face of genetic augmentation technologies.

Emotional and Psychological Augmentation: Nurturing Mental Well-Being

Space colonization recognizes the importance of emotional and psychological well-being in the face of isolation and cosmic vastness. This section explores emotional augmentation technologies, including AI-driven companions, virtual reality therapy, and neural modulation techniques. AI companions provide companionship and emotional support, mitigating feelings of loneliness among colonists. Virtual reality therapy offers immersive experiences, helping individuals cope with stress, anxiety, and homesickness. Neural modulation techniques, such as transcranial magnetic stimulation, alleviate symptoms of depression and enhance emotional resilience. Ethical considerations revolve around the authenticity of emotions, the impact on interpersonal relationships, and the potential blurring of lines between real and artificial experiences.

Ethical Concerns in Human Augmentation: Navigating Moral Frontiers

The integration of human augmentation technologies raises profound ethical concerns within space colonies. This section delves into the moral complexities surrounding consent, autonomy, and the implications of irreversible augmentations. Colonists face decisions about the enhancement of their biological selves, navigating the balance between individual choice and societal expectations. Ethical frameworks are established to safeguard personal liberties while ensuring the responsible use of augmentation technologies. Societal attitudes toward augmented individuals shape the inclusivity of space colonies, prompting reflections on discrimination, equality, and the fundamental rights of augmented persons.

Human-AI Integration: Symbiotic Partnerships in Space

Human-augmented individuals form symbiotic partnerships with artificial intelligence, creating a new paradigm of collaboration. This section explores the integration of AI-driven neural networks with human cognition, enabling enhanced problem-solving abilities and data analysis. Augmented astronauts collaborate with AI companions, fostering a synergy of human creativity and AI efficiency. Ethical considerations encompass questions of privacy, the integrity of thought, and the potential for AI influence on human decision-making. The boundaries between human consciousness and artificial intelligence blur, raising profound philosophical and existential questions about the nature of intelligence and personhood.

The Future of Human Augmentation in Space Colonies: Ethical Imperatives

The chapter concludes by contemplating the future trajectory of human augmentation in space colonies, emphasizing ethical imperatives that guide the responsible advancement of augmentation technologies. Colonists and scientists grapple with questions about the long-term impact of augmentations on human evolution, societal cohesion, and the preservation of human identity. Ethical frameworks are continuously adapted, ensuring that human augmentation technologies align with the values of spacefaring civilizations. The chapter leaves readers with a profound reflection on the intersection of humanity, technology, and ethics, shaping the future of augmented existence in the cosmic frontier.

Chapter 30: Humanity's Cosmic Destiny

In the grand tapestry of the cosmos, humanity stands at the threshold of an awe-inspiring destiny, one that transcends the boundaries of Earth and extends into the boundless expanse of space. This chapter embarks on a contemplative journey, envisioning the myriad possibilities that await humanity as a spacefaring species, exploring interstellar colonization, societal evolution, and the profound collaborations with extraterrestrial civilizations that could shape the enduring odyssey of humankind.

Interstellar Colonization: The Migration to Distant Worlds

Interstellar colonization emerges as the pinnacle of human ambition. This section delves into the hypothetical scenarios of humanity's migration to distant exoplanets, exploring the challenges and innovations that pave the way for interstellar travel. Generation ships, propelled by advanced propulsion systems and supported by self-sustaining ecosystems, embark on journeys spanning generations. Cryogenic hibernation technologies ensure the preservation of colonists during millennia-long voyages. Terraforming methods, including atmospheric manipulation and bioengineering, transform alien worlds into habitable environments. Ethical considerations surface regarding the preservation of planetary ecosystems and the rights of potential indigenous life forms, sparking debates about the responsibilities of humanity as cosmic pioneers.

The Evolution of Space Societies: Adapting to Extraterrestrial Realities

Space societies undergo profound evolution in the face of extraterrestrial challenges. This section explores the cultural, political, and economic transformations within spacefaring civilizations. Cultural syncretism flourishes as diverse human cultures merge and adapt to cosmic environments, fostering a tapestry of traditions and beliefs. Political systems grapple with questions of governance, representation, and decision-making across vast interstellar distances. Economic models explore resource allocation, trade networks, and the establishment of interstellar markets. Space societies embrace diversity, inclusivity, and cooperation, transcending national borders and Earthly divisions, fostering a sense of shared cosmic identity among colonists.

First Contact Scenarios: Bridging Humanity and Extraterrestrial Life

The prospect of first contact with extraterrestrial civilizations captivates the human imagination.

This section explores the diverse scenarios of inter-species communication, from subtle interstellar signals to face-to-face encounters with alien species. Humans establish protocols for deciphering extraterrestrial languages, deciphering alien technologies, and understanding the cultural nuances of extraterrestrial societies. The psychological and societal impact of first contact reverberates through human societies, raising questions about xenophobia, diplomacy, and the preservation of human identity in the face of cosmic diversity. Humanity grapples with existential questions about its place in the universe, challenging established belief systems and igniting philosophical debates about the nature of intelligence and consciousness.

Collaboration with Extraterrestrial Civilizations: Synergies of Knowledge and Culture

Collaboration between humans and extraterrestrial civilizations becomes a cornerstone of interstellar relations. This section delves into the synergies of knowledge exchange, cultural integration, and mutual technological advancements that emerge from inter-species collaboration. Humans share their scientific discoveries, artistic expressions, and philosophical insights, enriching the collective wisdom of cosmic civilizations. Extraterrestrial cultures introduce humanity to new forms of art, advanced technologies, and unique perspectives on the universe. Collaborative ventures span scientific research, interstellar trade, and joint explorations of cosmic phenomena. Ethical considerations revolve around respecting the autonomy of extraterrestrial civilizations, acknowledging their cultural heritage, and fostering inter-species harmony based on equality and mutual respect.

Cosmic Ethics: Moral Frameworks for Interstellar Existence

As humanity ventures into the cosmic arena, ethical frameworks adapt to the complexities of interstellar existence. This section explores cosmic ethics, addressing issues such as environmental stewardship, the rights of sentient beings, and the preservation of cosmic heritage. Colonists establish ethical guidelines for planetary exploration, ensuring the protection of alien ecosystems and the responsible utilization of planetary resources. Cosmic ethics extend to sentient beings encountered in space, promoting empathy, cooperation, and non-interference in the development of extraterrestrial civilizations. The preservation of cosmic heritage becomes a moral imperative, prompting humans to protect ancient celestial artifacts and natural wonders, fostering a sense of reverence for the cosmic tapestry that envelops them.

The Legacy of Human Exploration: Inspiring Future Generations

The legacy of human exploration echoes through the corridors of space colonies, inspiring future generations to reach for the stars. This section delves into the enduring impact of cosmic pioneers on the collective human psyche. Space colonies become educational hubs, nurturing the curiosity of young minds and fostering a passion for exploration and discovery. The stories of interstellar voyages, first contact experiences, and cosmic collaborations become an integral part of human culture, shaping literature, art, and scientific inquiry. Human societies draw inspiration from the resilience, creativity, and boundless curiosity of cosmic explorers, galvanizing them to embark on new interstellar journeys, further expanding the horizons of human knowledge and understanding.

The Cosmic Tapestry: Humanity's Enduring Odyssey

The chapter culminates in a contemplation of the cosmic tapestry, where humanity's enduring odyssey unfolds against the backdrop of galaxies, nebulae, and celestial wonders. This section explores the profound sense of awe and wonder that cosmic exploration instills in the human spirit. Humans gaze at the night sky, contemplating their place in the vastness of the universe, marveling at the beauty and complexity of cosmic phenomena. The cosmic tapestry becomes a canvas for human imagination, a source of artistic inspiration, and a reminder of the boundless possibilities that await in the uncharted depths of space. The chapter leaves readers with a sense of profound humility, curiosity, and determination, encouraging them to continue the exploration of the cosmic frontier, embracing the infinite potential that lies within the human spirit.

Chapter 31: The Ethical Dimensions of Space Exploration

In the limitless expanse of the cosmos, humanity's foray into space raises profound ethical questions that demand careful consideration. This chapter delves into the intricate ethical dimensions of space exploration, exploring issues related to environmental preservation, cultural heritage, scientific responsibility, and the rights of potential extraterrestrial life forms. As humanity ventures deeper into the cosmic frontier, ethical frameworks must evolve to ensure the responsible and respectful exploration of the universe.

Preserving Cosmic Environments: Guardianship of Celestial Bodies

Space exploration necessitates a commitment to preserving celestial environments. This section examines the ethical imperatives of protecting extraterrestrial landscapes, celestial bodies, and cosmic phenomena. Protocols are established to prevent contamination, preserve natural formations, and respect the pristine nature of celestial environments. Ethical considerations delve into the definition of contamination, potential harm to extraterrestrial ecosystems, and the long-term impact of human activities on the cosmic landscape. Spacefaring civilizations adopt a stance of stewardship, becoming guardians of celestial bodies and promoting the responsible exploration of cosmic environments.

Cultural Heritage in Space: Respect for Extraterrestrial Artifacts

As humanity explores space, it encounters ancient extraterrestrial artifacts and structures, raising ethical questions about cultural heritage. This section explores the preservation of alien archaeological sites, artworks, and cultural remnants. Ethical guidelines are established to ensure the respectful study of extraterrestrial cultures, avoiding exploitation and cultural appropriation. Collaborative initiatives involve interdisciplinary teams of scientists, historians, and cultural experts to decipher the meanings and significance of extraterrestrial artifacts. Ethical dilemmas arise concerning the public dissemination of knowledge, balancing scientific curiosity with cultural sensitivity, and acknowledging the rights of potential alien civilizations to their cultural heritage.

Scientific Responsibility: Ethical Conduct in Cosmic Research

The pursuit of scientific knowledge in space exploration demands ethical responsibility. This section delves into the ethical conduct of scientific research in cosmic environments, emphasizing transparency, peer review, and the open sharing of data. Ethical guidelines are established to prevent biased research, misinformation, and the misinterpretation of cosmic phenomena. Space scientists grapple with questions of intellectual property, acknowledging the collaborative nature of cosmic research while respecting individual contributions. Ethical considerations extend to the responsible communication of scientific discoveries to the public, fostering scientific literacy, and ensuring that cosmic knowledge benefits humanity as a whole.

First Contact Protocols: Ethical Diplomacy with Extraterrestrial Life

The prospect of first contact with extraterrestrial civilizations requires meticulous preparation and ethical diplomacy. This section explores the formulation of first contact protocols, emphasizing principles of respect, non-interference, and cultural sensitivity. Ethical frameworks guide interactions with alien species, ensuring that humanity approaches first contact with humility and a willingness to learn. Intergalactic diplomacy involves linguistic analysis, cultural anthropology, and interdisciplinary collaborations to establish meaningful communication channels. Ethical dilemmas emerge concerning the potential impact of human actions on extraterrestrial societies, raising questions about the unintended consequences of contact and the ethical responsibilities of spacefaring civilizations.

Rights of Potential Extraterrestrial Life: Ethical Considerations

The existence of potential extraterrestrial life forms raises profound ethical questions regarding their rights and ethical treatment. This section explores ethical considerations surrounding the rights of alien life, addressing questions of sentience, consciousness, and moral agency. Ethical frameworks are established to prevent harm to potential alien life forms, ensuring that scientific exploration does not inadvertently cause harm or disruption to extraterrestrial ecosystems. Ethical debates revolve around questions of moral obligations to alien life forms, the definition of life, and the moral responsibilities of spacefaring civilizations as cosmic neighbors.

Space Colonization and Social Justice: Ensuring Inclusivity

As humanity establishes colonies in space, social justice and inclusivity become paramount ethical concerns. This section explores ethical considerations related to equitable resource distribution, representation, and the elimination of discrimination in cosmic societies. Space colonies implement policies to promote diversity and inclusivity, ensuring equal opportunities for all colonists regardless of nationality, gender, or background. Ethical dilemmas arise concerning the allocation of resources, addressing historical injustices, and fostering a sense of belonging among diverse communities in space. Spacefaring civilizations actively engage in social justice initiatives, advocating for equality and social harmony in cosmic environments.

Ethics in Terraforming: Balancing Adaptation and Preservation

Terraforming alien worlds raises ethical questions about the intentional alteration of planetary environments. This section examines the ethical considerations of terraforming, balancing the adaptability of planets for human habitation with the preservation of existing ecosystems. Ethical frameworks guide terraforming projects, ensuring that native life forms, if present, are preserved and integrated into transformed environments. Ethical dilemmas emerge concerning the intentional modification of planetary atmospheres, geological structures, and climate patterns, raising questions about the moral responsibility of humanity as planetary engineers. Spacefaring civilizations engage in ethical debates, seeking consensus on the moral imperatives of terraforming and the preservation of planetary biodiversity.

The Ethical Imperative of Cosmic Exploration: Lessons for Humanity

The chapter concludes by reflecting on the ethical imperative of cosmic exploration, drawing lessons for humanity's journey into the stars. Ethical principles of respect, humility, cooperation, and environmental stewardship become guiding lights for spacefaring civilizations. Humanity acknowledges its place within the cosmic community, embracing the interconnectedness of all life in the universe. Ethical space exploration becomes a beacon of hope, inspiring future generations to explore the cosmos with a sense of wonder, reverence, and ethical responsibility. The chapter leaves readers with a profound reflection on the moral dimensions of space exploration, encouraging humanity to venture into the stars with a deep commitment to ethics and the preservation of cosmic harmony.

Chapter 32: Space Colonies and Cultural Adaptation

In the vast cosmic expanses of space colonies, human cultures undergo a profound metamorphosis, shaped by the unique challenges and opportunities of their extraterrestrial environments. This chapter delves into the intricate tapestry of cultural adaptation within these colonies, exploring the emergence of distinct languages, traditions, and belief systems influenced by the cosmic milieu. It examines the preservation of Earthly cultures within the colonies and the harmonious blending of diverse cultural elements, reflecting the kaleidoscope of human diversity in the cosmic expanse.

Cultural Synthesis in Space: Forging New Traditions

Space colonies become crucibles of cultural synthesis, where traditions from Earth meld with the realities of cosmic living. Here, diverse customs and practices intertwine, giving rise to vibrant new forms of expression. Colonists, hailing from various corners of Earth, intermingle and share their stories, culinary delights, and artistic creations. Hybrid languages emerge, blending elements from different Earth languages and incorporating space-related terminology. These languages become the linguistic heartbeat of space colonies, a testament to the adaptability and creativity of human communication.

Language Evolution in Microgravity: Cosmic Lexicons

Language in space undergoes a fascinating evolution, adapting to the unique challenges of microgravity and the cosmic environment. Slang and technical jargon seamlessly integrate into everyday speech, reflecting the spacefaring lifestyles of colonists. Multilingualism flourishes, as inhabitants converse effortlessly in multiple languages, fostering rich cross-cultural dialogues. Cosmic languages naturally incorporate terms related to celestial phenomena, spacecraft technologies, and the wonders of the cosmos. These evolving lexicons echo the spirit of exploration, capturing the essence of humanity's cosmic journey.

Cosmic Milestones: Rituals and Celebrations

Space colonies cultivate unique rituals and traditions, marking significant cosmic milestones and fostering a sense of community among colonists. Cosmic festivals celebrate celestial events such as eclipses, meteor showers, and planetary alignments, bringing colonists together for communal stargazing and astronomical storytelling. Space mission ceremonies become cherished cultural events, where explorers are bid farewell with reverence and welcomed back as heroes upon their return. Interstellar discoveries inspire rituals of gratitude, emphasizing the awe and wonder evoked by the cosmic unknown.

These traditions form a cultural tapestry that binds colonists in a shared cosmic heritage.

Belief Systems in Space: Cosmic Spirituality and Ethical Frontiers

Within the cosmic expanse, belief systems take on new dimensions, adapting to the challenges of space exploration. Cosmic spirituality emerges, emphasizing the interconnectedness of all life in the universe and fostering a deep sense of cosmic unity. Ethical frameworks evolve, underscoring the importance of environmental stewardship, inclusivity, and the preservation of cosmic heritage. Religious practices from Earth find new expressions in space, incorporating space-related symbolism and cosmic narratives. Philosophical discourse becomes a cornerstone of cultural life, as inhabitants contemplate the mysteries of existence, consciousness, and humanity's role in the cosmos.

Preserving Earth's Legacy: Museums of Terrestrial Heritage

Space colonies become custodians of Earth's rich cultural heritage, preserving it for future generations. Museums, libraries, and digital archives within colonies showcase the diverse tapestry of human history, displaying artifacts, artworks, and literary masterpieces from Earth. Cultural exchange programs flourish, allowing colonists to share the traditions and languages of their homelands. These museums serve as bridges between Earth's past and humanity's cosmic future, ensuring that the legacy of Earth's cultures continues to inspire, educate, and unite spacefaring civilizations on their cosmic journey.

Chapter 33: Space Colonies and Disaster Management

Asteroid Impact Preparedness: Planetary Defense Strategies

In the face of potential asteroid impacts, space colonies employ advanced planetary defense strategies. Colonies invest in telescopic surveillance to detect near-Earth objects well in advance, allowing for timely assessments of impact risks. Protocols are established for diverting asteroids, involving spacecraft equipped with propulsion systems capable of altering the object's trajectory. Colonies also collaborate on international efforts to monitor and mitigate global asteroid threats, fostering a collective planetary defense network.

Solar Flare Resilience: Shielding Against Cosmic Radiation

Space colonies face the threat of solar flares, intense bursts of energy from the Sun that release harmful cosmic radiation. Colonies utilize innovative shielding technologies, including electromagnetic shields and thick radiation-resistant materials, to protect habitats and vital systems. Emergency protocols are in place to redirect inhabitants to shielded areas during solar flare warnings. Advanced monitoring satellites stationed closer to the Sun provide real-time data, allowing colonies to predict and prepare for incoming solar flares, ensuring the safety of colonists.

Technical Failure Contingency: Redundancy and Maintenance

Space colonies recognize the critical importance of technical systems in sustaining life. Rigorous redundancy is built into essential systems, such as life support, power generation, and communication. Colonies maintain a fleet of robotic repair drones equipped with AI systems, capable of diagnosing and fixing technical issues remotely. Regular maintenance schedules are meticulously followed, ensuring that all systems remain in optimal condition. Colonists are trained in basic technical skills, allowing for manual interventions in case of emergencies, reinforcing the resilience of the colony's infrastructure.

Pandemic Preparedness: Isolation Protocols and Medical Facilities

Space colonies implement stringent pandemic preparedness measures, considering the confined living spaces and potential for fast disease spread. Isolation protocols are established to quarantine infected individuals, and medical facilities are equipped with state-of-the-art diagnostic tools and treatments. Colonists receive comprehensive medical training, allowing them to respond effectively to health emergencies. Additionally, colonies maintain a robust stockpile of medical supplies, including vaccines and antiviral medications, ensuring the ability to address unforeseen health crises swiftly.

Resource Scarcity Solutions: Sustainable Resource Management

Space colonies employ sustainable resource management strategies to mitigate potential resource scarcity during disasters. Water recycling systems are optimized for maximum efficiency, ensuring minimal water wastage. Hydroponic farms and bioregenerative life support systems provide a renewable source of food and oxygen, reducing dependency on external supplies. Colonies invest in asteroid mining technologies, extracting essential minerals and metals from nearby celestial bodies, ensuring a steady supply of resources even in the face of scarcity on Earth.

Cybersecurity Measures: Protecting Against Digital Threats

As digital systems underpin every aspect of colony life, robust cybersecurity measures are paramount. Colonies employ advanced encryption algorithms, AI-driven intrusion detection systems, and blockchain technologies to safeguard data integrity and prevent cyber-attacks. Ethical hacking and regular security audits are conducted to identify vulnerabilities and reinforce the colony's digital defenses. Colonists are educated on cybersecurity best practices, creating a vigilant community that actively participates in maintaining the integrity of digital networks.

Psychological Resilience Programs: Mental Health Support

Recognizing the psychological challenges posed by the isolation of space, colonies establish comprehensive mental health support programs. Psychologists and counselors are integral members of the colony's medical staff, offering regular counseling sessions and workshops on stress management, resilience, and interpersonal relationships. Virtual reality simulations provide colonists with immersive experiences of Earth's natural environments, mitigating homesickness and promoting mental well-being. Community activities, artistic expression, and sports are encouraged, fostering a sense of belonging and camaraderie among colonists.

Inter-Colony Collaboration: Mutual Aid Agreements

Space colonies establish mutual aid agreements, forming a network of support in times of disaster. Agreements include protocols for resource sharing, emergency evacuations, and medical assistance. Regular joint training exercises are conducted between colonies, ensuring seamless coordination in the event of a disaster.

Communication protocols are standardized, allowing colonies to share critical information swiftly. This collaborative approach strengthens the resilience of individual colonies, creating a unified front against potential threats in the cosmic frontier.

Chapter 34: Space Colonies and Artificial Gravity

Rotating Habitats: Simulating Gravity through Centrifugal Force

Rotating habitats are pivotal in providing space colonies with artificial gravity. By rotating around a central axis, these habitats generate centrifugal force, effectively simulating gravity for inhabitants. The design of these habitats involves careful calculations to ensure the right balance between rotation speed and gravitational pull. Colonies utilize specialized engineering materials, lightweight yet robust, to construct rotating sections. The result is a habitable environment where occupants experience a consistent gravitational force, allowing for a semblance of normalcy in daily activities.

Health Benefits: Preventing Muscular Atrophy and Bone Density Loss

Artificial gravity significantly contributes to the health and well-being of colonists. In a microgravity environment, humans are prone to muscular atrophy and bone density loss due to lack of resistance against gravity. Rotating habitats counteract these effects by subjecting the body to a constant gravitational force. Regular activities within these habitats, such as walking and exercising, help maintain muscle mass and bone density. This preventive approach ensures that colonists remain physically fit and reduces the risk of health complications associated with extended periods in space.

Adaptation Challenges: Human Adjustments to Altered Gravity

While artificial gravity provides numerous health benefits, it also poses challenges related to human adaptation. Colonists transitioning between microgravity and artificial gravity environments may experience disorientation and balance issues. The vestibular system, responsible for balance, needs time to adjust to the changing gravitational conditions. Colonies implement gradual acclimatization programs, allowing newcomers to adapt slowly. These programs include exercises designed to strengthen the vestibular system, ensuring a smoother transition and reducing the risk of motion sickness and related discomfort.

Structural Engineering: Designing Rotating Habitats for Stability

The structural engineering of rotating habitats demands meticulous planning to ensure stability and safety. Architects and engineers employ advanced computer simulations to model stress distribution, determining the optimal shape and size of rotating sections. Materials with high tensile strength are chosen to withstand the forces exerted during rotation. Bearings and joints are meticulously engineered to minimize friction and ensure smooth rotation. Safety redundancies, including backup power systems and emergency braking mechanisms, are integrated into the design, providing fail-safes in case of unforeseen events.

Artificial Gravity Variability: Tailoring Gravitational Levels for Different Spaces

Space colonies recognize the need for adaptable gravitational levels to accommodate diverse activities. For example, residential areas may maintain Earth-like gravity to support comfortable living conditions. In contrast, research laboratories and manufacturing facilities might require lower gravity levels to conduct experiments unaffected by gravitational interference. Colonies design modular habitats where gravitational levels can be adjusted based on specific needs, enhancing the efficiency and effectiveness of various tasks.

Psychological Impact: The Influence of Gravity on Human Psyche

The presence of artificial gravity has profound psychological implications for colonists. Studies indicate that the perception of gravity positively influences mental well-being, creating a sense of grounding and stability. The psychological impact extends to the perception of time, with colonists reporting a more natural passage of time in environments with artificial gravity. This psychological stability contributes to a positive living experience, fostering mental health and interpersonal relationships within the colony.

Cultural Adaptation: Gravity as a Cultural Marker

The integration of artificial gravity becomes a cultural marker within space colonies. It shapes cultural norms and traditions, influencing art, dance, and even cuisine. Colonists develop unique customs centered around gravity, creating a sense of identity and community. Cultural events and celebrations often incorporate the experience of artificial gravity, further reinforcing its significance in the social fabric of the colony. Gravity, once a natural force, becomes a symbol of resilience, adaptability, and human ingenuity in the cosmic frontier.

Long-Term Viability: Ensuring Sustainable Artificial Gravity Solutions

The long-term viability of artificial gravity solutions is a focus of ongoing research within space colonies. Scientists and engineers continuously refine existing methods and explore innovative approaches to create stable, energy-efficient rotating habitats. Sustainability measures, such as energy recovery systems and regenerative braking, are implemented to minimize resource consumption..

Chapter 35: Space Colonies and Aquatic Habitats

Enclosed Water Environments: The Aquatic Oasis in Space

Aquatic habitats within space colonies represent a pioneering endeavor in creating sustainable ecosystems beyond Earth. Enclosed water environments, carefully designed and regulated, offer colonists access to aquatic life and the myriad benefits it provides. These habitats, equipped with advanced technology, mimic Earth's aquatic ecosystems, fostering an environment where marine life thrives. The integration of aquatic habitats transforms space colonies into multifaceted ecosystems, enhancing biodiversity and providing essential resources for colonists.

Role of Aquatic Life: Biodiversity and Ecological Balance

Aquatic life plays a crucial role in maintaining the ecological balance within space colonies. Fish, algae, and other aquatic organisms contribute to the recycling of nutrients, essential for the growth of plants and the overall health of the ecosystem. Additionally, they serve as a food source for colonists, providing a sustainable and diverse protein supply. The presence of aquatic life introduces natural biological processes, ensuring

285

the equilibrium of the colony's ecosystem and supporting the overall well-being of its inhabitants.

Water Filtration Systems: Engineering Clean and Sustainable Water

To sustain aquatic habitats, sophisticated water filtration systems are implemented within space colonies. These systems employ advanced filtration techniques, including biological, chemical, and physical processes, to purify water efficiently. Colonies utilize biofilters populated by beneficial bacteria that break down waste materials, ensuring water quality. Innovative technologies such as membrane filtration and ultraviolet disinfection further enhance the purification process, providing colonists with a continuous supply of clean, potable water. Water filtration systems are meticulously maintained and monitored to guarantee the sustainability of aquatic habitats.

Psychological Benefits: The Therapeutic Influence of Water

Aquatic habitats offer psychological benefits to colonists, contributing to mental well-being and stress reduction. The calming effect of water, combined with the mesmerizing movements of aquatic life, creates a tranquil atmosphere within the colony. Observation decks overlooking these aquatic environments become spaces for relaxation and contemplation, providing a respite from the challenges of space life. The presence of water-based environments promotes a sense of connection with nature, fostering a positive psychological environment and enhancing the overall quality of life for colonists.

Educational Significance: Aquatic Habitats as Learning Centers

Aquatic habitats serve as educational hubs within space colonies, offering unique learning opportunities for residents, especially children. Educational programs centered around marine biology, ecology, and aquaculture allow colonists to engage with and understand the intricacies of aquatic ecosystems. Interactive exhibits, workshops, and guided tours provide hands-on experiences, fostering scientific curiosity and environmental awareness. The educational significance of aquatic habitats extends to research initiatives, where scientists conduct experiments and studies to unravel the mysteries of marine life in extraterrestrial environments.

Resource Utilization: Sustainable Practices and Circular Economy

Aquatic habitats contribute to sustainable resource utilization within space colonies. The waste generated by colonists becomes a valuable resource for the aquatic ecosystem. Organic waste is broken down by microorganisms and aquatic organisms, producing nutrients that support the growth of plants and algae. These organisms, in turn, serve as food for fish and other aquatic life. This circular economy model ensures efficient use of resources and minimizes waste, aligning with the principles of sustainability and environmental stewardship.

Aesthetic Design: Integrating Nature into Colony Architecture

The aesthetic design of aquatic habitats reflects a harmonious integration of nature into colony architecture. Transparent materials, such as reinforced glass, create immersive viewing panels that allow colonists to observe aquatic life in its natural habitat. The play of natural light and the vibrant colors of marine life enhance the visual appeal of these habitats, transforming them into artistic installations within the colony. Aesthetic considerations extend to the landscaping around aquatic environments, creating lush greenery and serene spaces that complement the aquatic oasis.

Future Expansion: Scaling Aquatic Habitats for Larger Colonies

As space colonies expand and accommodate larger populations, the scalability of aquatic habitats becomes a focal point of development. Engineers and architects collaborate to design modular aquatic habitats that can be expanded vertically and horizontally.

Chapter 36: Space Colonies and Intergenerational Challenges

Societal Dynamics: Navigating Multigenerational Communities

Intergenerational space colonies bring forth complex societal dynamics as residents from different age groups coexist within the confines of the colony. These dynamics encompass social hierarchies, communication gaps, and evolving cultural norms across generations. Colonists must navigate these intricacies, fostering mutual understanding and harmony. Challenges such as generational conflicts and the preservation of traditions present opportunities for innovative solutions, encouraging a cohesive community where wisdom from older generations coalesces with the fresh perspectives of the youth.

Education Strategies: Tailoring Learning for Multigenerational Residents

Education in intergenerational space colonies requires a tailored approach, accommodating diverse learning styles and age-specific needs. The curriculum must encompass not only foundational knowledge but also advanced subjects vital for the colony's development. Educational programs focus on cultivating critical thinking, problem-solving skills, and adaptability, preparing younger generations for the challenges of space life. Concurrently, lifelong learning initiatives engage older residents, ensuring they remain intellectually stimulated and actively contribute to the colony's intellectual capital.

Knowledge Transfer: Preserving Wisdom Across Generations

Preserving the accumulated wisdom of older generations is paramount in intergenerational colonies. Knowledge transfer initiatives utilize mentorship programs, oral traditions, and digital repositories to capture and disseminate the expertise of experienced colonists. Apprenticeships and hands-on learning experiences facilitate the transfer of practical skills, ensuring vital knowledge in fields such as agriculture, engineering, and medical sciences is passed down seamlessly. The preservation of historical narratives and cultural heritage further enriches the collective identity of the colony.

Long Lifespans: Adapting to Extended Human Lifecycles

In the context of space colonies, advancements in medical technology often result in significantly extended human lifespans. Consequently, the challenges of addressing age-related health concerns, maintaining mental acuity, and promoting overall well-being become focal points. Geriatric care and psychological support systems are meticulously developed, ensuring that elderly residents lead fulfilling lives. The integration of recreational activities, social engagements, and intellectual pursuits into the daily lives of seniors promotes active aging and enriches the social fabric of the colony.

Sustainable Practices: Ensuring Resource Resilience for Future Generations

Intergenerational space colonies necessitate a sustainable approach to resource management. Responsible consumption, recycling initiatives, and renewable energy sources form the cornerstone of resource resilience. Colonists employ circular economy models, where waste products are repurposed into valuable resources, minimizing the ecological impact. Sustainable agriculture, aquaculture, and closed-loop systems ensure a continuous supply of food and water, reducing the colony's dependence on external resources and fostering self-sufficiency for generations to come.

Interdisciplinary Collaboration: Fostering Cross-Generational Innovation

Collaboration across generations becomes a catalyst for innovation in intergenerational space colonies. Multidisciplinary projects that involve residents of varying ages encourage the exchange of ideas and perspectives. Scientific research, artistic endeavors, and technological innovations benefit from the diverse experiences and knowledge bases of the colony's inhabitants. This collaborative spirit nurtures creativity, fuels advancements, and reinforces the sense of community, laying the foundation for a prosperous and harmonious multigenerational society.

Ethical Considerations: Balancing Interests Across Lifetimes

Ethical dilemmas arise in intergenerational space colonies, especially concerning resource allocation, medical interventions, and access to opportunities over extended lifetimes. Striking a balance between the needs and aspirations of different age groups is a continuous endeavor. Ethical frameworks are established to ensure fairness, equity, and inclusivity across lifetimes. Deliberative forums, where residents participate in decision-making processes, become essential in addressing these challenges, fostering a sense of ownership and shared responsibility within the colony.

Legacy and Continuity: Shaping the Future Through Intergenerational Bonds

Intergenerational bonds in space colonies create a legacy of shared experiences and collective progress. Older generations contribute not only knowledge but also cultural heritage, shaping the identity of the colony.

Chapter 37: Space Colonies and Cosmic Phenomena

Black Holes: The Enigmatic Vortices of Space

Black holes, mysterious cosmic entities with gravitational forces so intense that not even light can escape, pose both scientific intrigue and potential hazards to space colonies. Scientists within colonies study these enigmatic vortices to unravel the secrets of spacetime. The extreme conditions near black holes provide unique opportunities for testing theories in physics, including those related to quantum gravity. However, the proximity of a black hole demands rigorous safety protocols, ensuring that colonists are shielded from dangerous gravitational effects and high-energy radiation.

Supernovae: Celestial Explosions and Stellar Evolution

Supernovae, the explosive deaths of massive stars, enrich the cosmos with heavy elements and shape the formation of galaxies. Studying supernovae up close offers invaluable insights into stellar evolution and nucleosynthesis processes. Space colonies positioned strategically can observe supernovae events in neighboring star systems, enabling astronomers to monitor the intricate phases of these cosmic explosions. Safeguarding colonies from potential radiation bursts accompanying supernovae is crucial, necessitating advanced shielding technologies and early warning systems.

Cosmic Rays: High-Energy Particles from Distant Sources

Cosmic rays, high-energy particles originating from astrophysical sources such as supernovae remnants and active galactic nuclei, constantly bombard space colonies. Scientists utilize these particles to probe the fundamental nature of matter and the universe. Cosmic ray detectors within colonies capture and analyze these particles, unveiling clues about the origins of cosmic rays and the cosmic accelerators responsible for their immense energies. Despite their scientific significance, protective shielding is essential to safeguard colonists from the harmful effects of cosmic rays, ensuring their well-being.

Gamma-Ray Bursts: Cosmic Explosions and High-Energy Radiation

Gamma-ray bursts, the most energetic events in the universe, emit powerful bursts of gamma rays during stellar cataclysms like supernovae or neutron star mergers. These bursts, detectable from vast distances, provide astronomers with valuable data on the early universe and extreme astrophysical phenomena. Space colonies equipped with gamma-ray detectors contribute to the understanding of these events, unraveling their origins and implications. However, careful planning is paramount to shield colonists from the intense radiation associated with gamma-ray bursts, mitigating potential health risks.

Neutrinos: Ghostly Particles from Cosmic Sources

Neutrinos, nearly massless particles produced in various cosmic processes, offer a unique window into astrophysical phenomena. Space colonies equipped with neutrino observatories detect these elusive particles, providing insights into processes like nuclear fusion in stars and supernova explosions. Neutrino studies within colonies enhance our understanding of the universe's fundamental interactions. Despite their weak interaction with matter, robust detection systems are crucial to capture these ghostly particles, enabling detailed analyses and advancing our knowledge of cosmic phenomena.

Gravitational Waves: Ripples in Spacetime

Gravitational waves, ripples in spacetime caused by cataclysmic events such as mergers of black holes or neutron stars, offer a revolutionary means of observing the universe. Space colonies housing gravitational wave detectors contribute to the global network of observatories, enhancing the sensitivity of gravitational wave detections. Studying these waves provides valuable data on the properties of celestial objects and tests Einstein's theory of general relativity. Careful calibration and maintenance of detectors are essential to capture faint gravitational wave signals, ensuring the accuracy of scientific observations.

Magnetars: Intensely Magnetized Neutron Stars

Magnetars, neutron stars with exceptionally strong magnetic fields, exhibit bursts of X-rays and gamma rays, making them intriguing cosmic phenomena. Studying magnetars in proximity to space colonies sheds light on the extreme magnetic and gravitational conditions near these celestial objects. Observations help scientists understand the origin of their powerful magnetic fields and their influence on the surrounding interstellar medium. Colonies, equipped with specialized telescopes, provide a platform for continuous monitoring of magnetar activities, enriching our knowledge of these enigmatic objects.

Pulsars: Cosmic Lighthouses and Stellar Clocks

Pulsars, rotating neutron stars emitting beams of electromagnetic radiation, serve as precise celestial clocks and cosmic laboratories. Space colonies housing pulsar observation facilities study these rapidly spinning remnants of supernovae, unraveling their properties and aiding in tests of gravitational theories. Pulsar timing arrays within colonies contribute to the detection of low-frequency gravitational waves, opening new avenues for gravitational wave astronomy. Maintaining stable observational platforms and data analysis systems is essential for accurate pulsar studies, enabling detailed investigations into the universe's fundamental physics.

Chapter 38: Space Colonies and Virtual Reality

Virtual Reality in Astronaut Training: Simulating Space Environments

In space colonies, virtual reality (VR) serves as a pivotal tool for training astronauts. Sophisticated VR simulations recreate space environments, allowing astronauts to practice complex tasks, conduct extravehicular activities, and troubleshoot emergencies. These simulations provide a safe yet realistic platform, ensuring that astronauts are well-prepared for the challenges of space missions. VR training modules enable continuous skill enhancement, fostering a highly trained and adaptable spacefaring workforce.

Telepresence Robotics: Remote Operations with Precision

Telepresence robotics, enabled by VR technology, revolutionize space operations. Remote-controlled robots equipped with VR interfaces allow precise manipulations and intricate repairs in space colonies. Astronauts can operate these robots from a distance, performing tasks that require dexterity and attention to detail. This advancement enhances the efficiency of maintenance and scientific experiments, minimizing the risks associated with extravehicular activities and ensuring the optimal functioning of colony systems.

VR-Based Scientific Research: Exploring Uncharted Territories

VR facilitates immersive scientific research within space colonies. Scientists can delve into complex datasets, visualize molecular structures, and explore astronomical phenomena through VR interfaces. This interactive approach accelerates scientific discovery by enabling researchers to analyze data in three-dimensional space, fostering new insights and discoveries. VR-based research environments enhance collaboration among scientists, encouraging interdisciplinary studies and innovative problem-solving approaches.

Recreational VR Spaces: Escaping to Virtual Worlds

In the isolated environment of space colonies, recreational VR spaces provide essential outlets for relaxation and entertainment. Colonists can escape the confines of their immediate surroundings by immersing themselves in virtual worlds, engaging in interactive games, artistic experiences, or virtual tourism. These recreational activities promote mental well-being, combatting feelings of isolation and monotony. VR-based social platforms enable colonists to connect with others, fostering a sense of community despite physical distance.

Psychological Benefits of VR: Combating Isolation and Enhancing Social Interactions

VR technology plays a crucial role in maintaining colonists' psychological well-being. Immersive VR experiences offer a sense of presence and connection, mitigating feelings of isolation and confinement. Virtual social spaces allow colonists to interact with friends, family, and fellow colonists in lifelike environments, strengthening social bonds and reducing the psychological challenges of long-term space habitation. VR therapy sessions and relaxation programs further contribute to mental health, ensuring the resilience and adaptability of colonists in the face of isolation-related stressors.

Educational VR Platforms: Enhancing Learning Experiences

In space colonies, educational VR platforms revolutionize learning experiences. Virtual classrooms equipped with interactive simulations enable engaging lessons on diverse subjects, from astronomy to biology. Students can explore virtual laboratories, conduct experiments, and interact with historical events, enhancing their understanding through immersive learning. VR-based education fosters curiosity, critical thinking, and adaptability, preparing future generations of colonists for the challenges and opportunities of space exploration.

VR-Based Design and Planning: Optimizing Colony Layouts

VR technology aids architects and engineers in designing and planning space colony layouts. Virtual architectural simulations allow professionals to visualize and optimize living spaces, work environments, and recreational areas. This iterative design process ensures efficient space utilization, ergonomic considerations, and aesthetic harmony within colonies. VR-based planning enhances the functionality and aesthetics of space habitats, creating environments that optimize both practicality and residents' quality of life.

VR-Based Communication with Earth: Bridging Interstellar Distances

VR-based communication systems bridge the vast interstellar distances between space colonies and Earth. Immersive VR environments enable real-time communication with loved ones on Earth, replicating face-to-face interactions. These emotionally rich connections foster a sense of belonging and maintain ties with Earth's culture and society. VR communication also serves educational and cultural exchange purposes, enabling colonists to participate in global events and initiatives, ensuring that the cosmic pioneers remain connected to their terrestrial roots.

Chapter 39: Space Colonies and Language Evolution

The Genesis of Space Linguistics: Adapting Language to Extraterrestrial Life

Within the unique confines of space colonies, languages undergo a fascinating transformation. New terminologies emerge, shaped by the demands of spacefaring life. Concepts related to life support systems, extraterrestrial agriculture, and advanced technology necessitate the creation of novel vocabulary. Space linguists analyze the linguistic needs of colonists, inventing words and phrases that encapsulate the intricacies of space living. This linguistic evolution reflects the adaptability of human communication to unprecedented environments.

Linguistic Adaptations to Space Environment: The Influence of Microgravity

Microgravity significantly impacts language use within space colonies. Linguistic adaptations occur to convey experiences unique to the absence of gravity. Words and expressions related to movement, stability, and spatial orientation take on new meanings. Space colonists develop spatial language to navigate their environment effectively, ensuring clear communication amid weightless conditions. This linguistic adaptation showcases humanity's ability to adapt language to the physical challenges of space habitation.

Cultural Influences on Language: A Tapestry of Multilingualism

Space colonies, comprising individuals from diverse cultural backgrounds, experience a rich interplay of languages. Multilingualism becomes a cornerstone of communication, fostering an environment where cultural exchange shapes linguistic evolution. Phrases, idioms, and pronunciation patterns from various Earth languages blend and enrich the linguistic landscape of colonies. This multicultural linguistic tapestry highlights the harmonious coexistence of cultures within space habitats, emphasizing the shared human experience that transcends language barriers.

Interstellar Communication: Language as a Universal Bridge

The need for communication with Earth and potential extraterrestrial civilizations necessitates the development of universal languages within space colonies. Colonists work collectively to create languages that bridge linguistic divides, ensuring efficient communication across cultures and generations. These constructed languages, known as "space lingua," draw inspiration from Earth languages while incorporating simplified grammar and universal vocabulary. Space lingua serves as a medium for interstellar communication, uniting humanity in the cosmic quest for knowledge and understanding.

Language Preservation and Cultural Identity

Amid the influx of diverse languages, efforts are made to preserve Earth languages within space colonies. Language preservation initiatives celebrate cultural heritage, ensuring that traditional languages are passed down to future generations. Cultural events, storytelling sessions, and language classes are organized to maintain linguistic diversity. Preserving native languages fosters a sense of identity, connecting colonists to their roots and heritage, even in the distant realms of space.

Technological Advancements in Language Processing

In space colonies, advanced technology aids language processing and translation. AI-driven language tools enable seamless communication between multilingual colonists. Real-time translation devices, neural language interfaces, and immersive language learning programs enhance language skills and facilitate effortless conversations. These technological advancements break down language barriers, fostering collaboration and mutual understanding among inhabitants.

Language and Social Cohesion: The Role of Communication in Community Building

Language plays a pivotal role in building social cohesion within space colonies. Effective communication enhances collaboration, teamwork, and a sense of belonging among colonists. Multilingual communities celebrate language diversity, organizing language festivals and cultural exchanges. Clear communication channels, facilitated by language, ensure that ideas, emotions, and concerns are shared openly, fostering a supportive and tightly-knit community. Language becomes a powerful tool for nurturing social bonds and collective well-being in the isolated environment of space.

The Future of Space Linguistics: Adapting to Interstellar Challenges

As humanity ventures further into the cosmos, space linguistics continues to evolve. Colonists anticipate encounters with extraterrestrial life forms and the need for interstellar communication. Space linguists work on deciphering potential alien languages, developing protocols for communication with unknown civilizations. The future of space linguistics holds the promise of unraveling the intricacies of interstellar languages, fostering intergalactic dialogue, and expanding the horizons of human communication beyond the stars.

Chapter 40: Space Colonies and the Legacy of Humanity

Advancements in Scientific Understanding: Space as a Catalyst for Knowledge

The exploration of space has propelled humanity into an era of unprecedented scientific discovery. Space colonies serve as laboratories for cutting-edge research, offering insights into fields such as astronomy, physics, and biology. Scientists within colonies conduct experiments in microgravity, unravel cosmic mysteries, and study the behavior of materials in space. These discoveries not only enhance our understanding of the universe but also contribute to technological innovations on Earth, leaving an indelible mark on human knowledge.

Technological Marvels: Innovations Driven by Space Colonies

The challenges of space habitation foster technological innovation, leading to the creation of groundbreaking technologies. Advanced life support systems, efficient waste recycling methods, and sustainable energy solutions are developed to sustain life in the isolated environment of colonies. These innovations find applications on Earth, revolutionizing industries and improving the quality of life. Space-driven inventions, from medical devices to environmental monitoring systems, enrich human civilization, shaping a future where space technology benefits every corner of our planet.

Humanity's Curiosity Unleashed: The Quest for Understanding

At the heart of space exploration lies the enduring human spirit of curiosity. Space colonies embody humanity's relentless pursuit of knowledge and the desire to unravel the mysteries of the cosmos. The insatiable curiosity of scientists, engineers, and explorers within colonies drives them to push the boundaries of what is known. This innate thirst for understanding fuels the exploration of distant planets, the study of celestial phenomena, and the search for extraterrestrial life. The legacy of this curiosity becomes a beacon, inspiring future generations to explore, question, and expand the horizons of human understanding.

Cultural and Artistic Expressions: Space as a Muse

The cosmic expanse serves as a boundless muse for artists, writers, and creators within space colonies. Colonists channel their experiences, emotions, and awe of the universe into artistic expressions. Space-inspired artworks, literature, music, and performances capture the imagination of humanity, bridging the gap between the cosmos and Earth. These cultural creations become a testament to the profound impact of space exploration on human creativity, evoking a sense of wonder and inspiring generations to dream beyond the stars.

Preserving the Ethical and Moral Compass: Lessons from Space

In the pursuit of space colonization, humanity faces ethical dilemmas and moral challenges. Space colonies serve as crucibles for testing ethical frameworks, exploring questions related to genetic engineering, resource allocation, and societal harmony. Ethical decisions made within colonies shape the moral compass of humanity's cosmic journey. Lessons learned from navigating these dilemmas contribute to the evolution of ethical principles on Earth, guiding humanity toward responsible scientific practices, social justice, and environmental stewardship. The legacy of ethical considerations becomes an integral part of humanity's enduring journey into the cosmos, ensuring that our exploration of space is grounded in ethical integrity and compassion.

Chapter 41: Space Colonies and Cosmic Research

Stargazing Beyond Earth's Atmosphere: Astronomical Observations in Space Colonies

The absence of Earth's atmosphere and light pollution in space colonies creates an ideal environment for astronomical observations. Telescopes and observation instruments are deployed to study celestial objects with unprecedented clarity. Astronomers within colonies explore distant stars, galaxies, and nebulae, providing invaluable data that enhances our understanding of the universe. These observations contribute to discoveries related to cosmic phenomena, stellar evolution, and the formation of galaxies, enriching humanity's knowledge of the cosmos.

Particle Physics in Microgravity: Probing the Fundamentals of Matter

Space colonies serve as unique laboratories for particle physics experiments. In the microgravity environment, scientists conduct experiments to study the behavior of subatomic particles. Particle accelerators within colonies enable researchers to explore fundamental questions about the nature of matter and the universe. By observing particle interactions in space, scientists gain insights into the building blocks of the universe, contributing to the advancement of theoretical physics and our understanding of the fundamental forces governing the cosmos.

Cosmological Studies: Understanding the Universe's Evolution

The vastness of space within colonies allows for extensive cosmological studies. Scientists study cosmic microwave background radiation, cosmic rays, and gravitational waves to unravel the secrets of the universe's evolution. Cosmologists within colonies explore the Big Bang theory, cosmic inflation, and the structure of spacetime, providing critical data to refine cosmological models. These studies offer glimpses into the universe's origins, shedding light on its past, present, and future, shaping our understanding of the cosmic timeline.

Dark Matter and Dark Energy: Probing the Cosmic Mysteries

One of the most significant cosmic mysteries lies in the nature of dark matter and dark energy. Space colonies facilitate experiments to detect and study these enigmatic components of the universe. Scientists within colonies employ sophisticated detectors and sensors to search for dark matter particles and measure the effects of dark energy on cosmic expansion. These experiments aim to unravel the mysteries behind the accelerated expansion of the universe and the unseen forces shaping cosmic structures, offering profound insights into the nature of the cosmos.

Astrobiology: Exploring Extraterrestrial Life

Space colonies provide a platform for astrobiological research, aiming to answer the age-old question: "Are we alone in the universe?" Scientists study extremophiles, organisms capable of surviving in extreme conditions, to understand the potential for life on other celestial bodies. Colonies conduct experiments to simulate extraterrestrial environments, exploring the possibility of life on Mars, Europa, and other celestial bodies. These investigations contribute to humanity's search for extraterrestrial life, offering clues about the conditions necessary for life to thrive beyond Earth.

Chapter 42: Space Colonies and Cosmic Exploration

Interplanetary Missions: Probing the Secrets of Nearby Worlds

Space colonies serve as mission control centers for interplanetary exploration. Robotic missions are dispatched to nearby planets such as Mars, Venus, and the gas giants like Jupiter and Saturn. These missions involve landers, rovers, and orbiters equipped with advanced scientific instruments to study the geology, atmosphere, and potential habitability of these celestial bodies. Data collected from these missions provide crucial insights into the history and composition of neighboring planets, offering valuable comparative data for Earth's geological studies.

Asteroid Mining: Harvesting Resources Beyond Earth

Space colonies pioneer asteroid mining operations, tapping into the vast wealth of resources found within asteroids. Advanced spacecraft equipped with mining equipment are dispatched to asteroids rich in minerals, metals, and water ice. Colonists develop techniques to extract and process these resources, addressing Earth's resource scarcity and fueling the growth of space industries. Asteroid mining also plays a pivotal role in supporting the sustainability of space colonies by providing essential materials for construction, manufacturing, and life support systems.

Intergalactic Probes: Exploring the Depths of Interstellar Space

Space colonies embark on ambitious missions to explore interstellar space. Deep space probes equipped with powerful telescopes and scientific instruments are launched toward the boundaries of our solar system and beyond. These probes study the interstellar medium, cosmic rays, and magnetic fields, shedding light on the interstellar environment. Data collected by these probes contribute to our understanding of the universe's vastness, helping scientists comprehend the dynamics of space beyond the confines of our solar neighborhood.

Wormholes and Faster-Than-Light Travel: Theoretical Explorations

Within space colonies, physicists and engineers theorize about the possibility of wormholes— hypothetical passages through spacetime that could allow for faster-than-light travel. Theoretical research explores the potential existence of stable wormholes and the technology required to create and navigate them. Scientists within colonies collaborate on experimental designs and simulations to test the feasibility of traversable wormholes, aiming to revolutionize space travel and unlock the mysteries of distant galaxies and exoplanetary systems.

Interstellar Communication: Reaching Out to Extraterrestrial Civilizations

Space colonies engage in interstellar communication efforts, sending intentional signals into space in the hopes of reaching extraterrestrial civilizations. Colonists collaborate on designing and transmitting interstellar messages that encapsulate humanity's knowledge, culture, and achievements. These transmissions, sent via powerful radio telescopes, serve as a testament to human curiosity and the desire to connect with potential extraterrestrial neighbors. Scientists within colonies also dedicate efforts to analyzing incoming signals, remaining vigilant for any signs of communication from civilizations beyond our solar system.

Chapter 43: Space Colonies and Cosmic Connection

The Cosmic Mind: Exploring Universal Consciousness

In the contemplative atmosphere of space colonies, residents delve into philosophical inquiries about the interconnectedness of all things in the universe. Concepts such as cosmic consciousness and the universal mind are explored, reflecting on the idea that human consciousness is not separate from the cosmos but rather an integral part of it. Philosophers and thinkers within colonies engage in deep discussions about the nature of existence, exploring the profound unity between human consciousness and the vast cosmic expanse.

Transcending Boundaries: Oneness with the Universe

Space colonies foster a sense of oneness with the universe among their inhabitants. Residents contemplate the notion of transcending individual identity and merging with the cosmic whole. Discussions delve into spiritual practices and meditative techniques that aim to dissolve the boundaries between self and the universe. The pursuit of oneness becomes a central theme, encouraging individuals to explore their spiritual selves and experience a profound connection with the cosmic energies that permeate space and time.

Stardust and Spirituality: Embracing the Essence of Creation

Within space colonies, dwellers ponder the ancient wisdom that claims humans are made of stardust, forged from the remnants of exploded stars. This realization fosters a deep spiritual connection to the universe. Residents explore spiritual rituals and ceremonies that honor the cosmic origins of life. Symbolic practices, such as stardust ceremonies, become part of the colony's cultural fabric, celebrating the interconnectedness of all living beings with the celestial forces that shaped the cosmos.

Existential Contemplations: Humanity's Place in the Cosmic Tapestry

In the quietude of space, inhabitants engage in existential contemplations about humanity's place in the cosmic tapestry. Philosophers, theologians, and scientists ponder questions about the purpose of life, the existence of extraterrestrial intelligence, and the ultimate fate of the universe. These discussions lead to diverse perspectives on the significance of human existence and our role in the grand cosmic narrative. Contemplating the vastness of the universe, residents grapple with the mysteries of existence, inspiring artistic expressions, philosophical treatises, and spiritual insights.

Harmony of Being: Aligning with Cosmic Energies

Space colonies become centers for exploring the harmony of being in alignment with cosmic energies. Inhabitants delve into practices that promote physical, mental, and spiritual well-being, drawing inspiration from cosmic principles. Yoga, meditation, and mindfulness techniques are embraced, allowing colonists to attune themselves to the rhythms of the universe. Holistic approaches to health and wellness become integral to colony life, emphasizing the interconnectedness of mind, body, and spirit with the cosmic energies that pervade space.

Chapter 44: Space Colonies and Cosmic Collaboration

Intergalactic Diplomacy: Forging Alliances Beyond the Stars

Space colonies become hubs of intergalactic diplomacy, engaging in dialogues with advanced civilizations from distant galaxies. Diplomats and scholars from colonies participate in intricate negotiations, fostering alliances, and promoting peaceful coexistence. Discussions encompass shared resources, scientific cooperation, and mutual defense strategies. The exchange of ambassadors becomes a common practice, marking a new era of collaboration that transcends the confines of individual star systems.

Cultural Fusion: Celebrating Diversity Across Cosmic Realms

The cultural exchange between space colonies and extraterrestrial civilizations enriches the tapestry of cosmic diversity. Colonists immerse themselves in the traditions, art forms, and languages of alien cultures, fostering mutual understanding and appreciation. Cultural festivals and events become forums for celebrating the harmonious coexistence of diverse cosmic societies. Through these interactions, space dwellers embrace the richness of interstellar cultures, fostering a sense of unity amidst their differences.

Technological Synergy: Advancing Civilization Through Knowledge Exchange

The collaborative efforts between space colonies and advanced civilizations lead to a profound exchange of technological expertise. Scientists and engineers share discoveries and innovations, accelerating the development of groundbreaking technologies. From energy sources harnessed from exotic cosmic phenomena to advancements in artificial intelligence and interstellar travel, the knowledge exchange fuels the progress of both cosmic societies. Joint research projects become common, driving the evolution of civilizations towards new frontiers of scientific understanding.

Cosmic Ethics: Exploring Ethical Frameworks Across Universes

In the realm of interstellar collaboration, discussions about ethics and morality become paramount. Scholars, philosophers, and spiritual leaders from diverse cosmic realms engage in dialogues about ethical frameworks that govern the behavior of intelligent civilizations. Deliberations revolve around topics such as sentient rights, environmental stewardship, and the preservation of cosmic heritage. The exchange of ethical perspectives fosters a shared commitment to universal values, laying the foundation for ethical cooperation and mutual respect among cosmic societies.

Guardians of Cosmic Harmony: Preserving Peace in the Vastness of Space

Space colonies, in collaboration with advanced civilizations, take on the responsibility of preserving peace in the cosmic expanse. Jointly established intergalactic peacekeeping forces work tirelessly to mediate conflicts, prevent interstellar wars, and maintain stability across star systems. The commitment to cosmic harmony becomes a shared mission, transcending planetary boundaries and fostering a sense of interconnected security. Through cooperative efforts, spacefaring civilizations stand as guardians, ensuring that the vastness of space remains a realm of peaceful coexistence and boundless exploration.

Chapter 45: Space Colonies and Cosmic Wonder

Eternal Marvels: A Journey Through the Cosmic Spectacle

Space colonies become observatories to the universe's grandeur, allowing inhabitants to witness awe-inspiring phenomena. From the birth of stars in distant nebulae to the cataclysmic explosions of supernovae, colonists experience the eternal marvels that shape the cosmos. Telescopes and advanced sensors capture the dance of cosmic bodies, deepening humanity's understanding of the universe's vastness and beauty.

Cosmic Harmony: Finding Unity in the Celestial Orchestra

The intricate dance of celestial bodies within galaxies and cosmic clusters evokes a sense of cosmic harmony. Space colonies delve into the study of gravitational waves, cosmic radiation, and quantum interactions, unraveling the symphony that permeates the universe. Scientists and artists collaborate to interpret this cosmic harmony, exploring its resonance with human emotions and the interconnectedness of all things in the vast cosmic tapestry.

The Spirit of Exploration: Igniting Curiosity Across the Stars

Cosmic wonder fuels the spirit of exploration, inspiring spacefarers to venture into uncharted territories. Colonists embark on interstellar journeys, seeking out distant exoplanets, asteroid belts, and cosmic anomalies. The profound curiosity stirred by cosmic wonders drives the invention of advanced propulsion systems and spacecraft, enabling humanity to traverse cosmic distances previously deemed insurmountable.

Imagination Unleashed: Cosmic Inspiration in Art and Literature

Cosmic wonder serves as a wellspring of inspiration for artists and writers within space colonies. Poets pen verses that echo the serenity of interstellar voids, while painters capture the vibrant hues of distant nebulae on their canvases. Science fiction writers weave narratives of intergalactic adventures, exploring the mysteries of black holes and the possibility of extraterrestrial civilizations. The boundless canvas of the cosmos becomes a muse, fostering creativity that transcends the confines of Earth.

The Cosmic Connection: Spiritual and Existential Revelations

Beyond scientific inquiry, cosmic wonder leads to profound spiritual and existential revelations.

Space colonies become centers of contemplation, where philosophers and theologians explore the interconnectedness of all life forms and the universe. Spiritual seekers delve into cosmic consciousness, meditating on the cosmic energy that permeates existence. The eternal wonder of the cosmos becomes a catalyst for spiritual growth, nurturing a deep sense of reverence for the cosmic forces that shape reality.

Chapter 46: Space Colonies and the Infinite Cosmos

Cosmic Boundaries: Exploring the Limits of Space and Time

Space colonies serve as platforms for contemplating the cosmic boundaries that define the universe. Scientists and physicists delve into the concept of space-time, exploring the fabric of reality that weaves together galaxies, stars, and planets. They ponder the limits of the observable universe, questioning the nature of space beyond the reaches of our most powerful telescopes. In the boundless expanse of space colonies, humanity contemplates the unfathomable vastness that stretches beyond the cosmic horizon.

Multiverse Realities: Navigating Parallel Universes

Within the confines of space colonies, theoretical physicists delve into the intriguing realms of multiverse theory. They contemplate the existence of parallel universes, each with its own set of physical laws and cosmic constants. Scientists explore the mathematical intricacies that support the existence of multiple realities, challenging conventional notions of the universe's singularity. Colonists ponder the profound implications of a multiverse, considering the diversity of life, matter, and energy that could exist across alternate realities.

Quantum Mysteries: Embracing Uncertainty and Probability

Quantum mechanics, the enigmatic realm that governs the behavior of particles at the smallest scales, becomes a focal point of exploration within space colonies. Physicists grapple with the mysteries of quantum entanglement, superposition, and the probabilistic nature of subatomic particles. The inherent uncertainty of the quantum world challenges the very fabric of reality, prompting deep philosophical and existential inquiries. Inhabitants of space colonies delve into the nature of probability, questioning the deterministic foundations of the cosmos and embracing the inherent uncertainty that defines quantum phenomena.

The Cosmic Tapestry: Interconnectedness of All Realities

Space colonies become centers of contemplation on the interconnectedness of all realities. Scientists and philosophers explore the cosmic tapestry that binds together parallel universes, contemplating the threads that connect disparate realms of existence. They delve into the possibility of cosmic consciousness, where the universe itself is considered a sentient, interconnected entity. Within the contemplative atmosphere of space colonies, inhabitants ponder the profound implications of a cosmic tapestry that unites all realities, inviting exploration of the deepest mysteries of existence.

Existential Inquiry: Humanity's Place in the Infinite

In the vastness of the infinite cosmos, space colonies become havens for existential inquiry. Philosophers, theologians, and thinkers grapple with questions of human existence, purpose, and meaning in the face of an infinite and diverse cosmic landscape. Colonists contemplate the possibility of intelligent life in other universes, pondering the existential implications of cosmic loneliness or the potential for interconnectedness with extraterrestrial civilizations. Within the contemplative halls of space colonies, humanity embarks on a profound journey of self-discovery, seeking to understand its place in the infinite expanse of the cosmos.

Chapter 47: Space Colonies and Cosmic Enlightenment

The Quest for Cosmic Understanding: Exploring the Nature of Reality

Space colonies become sanctuaries for profound philosophical exploration, delving into the nature of reality itself. Philosophers and scholars contemplate metaphysical concepts, questioning the fabric of the universe and the essence of existence. Within the walls of space colonies, profound inquiries into the fundamental nature of reality ignite discussions, challenging conventional paradigms and inviting contemplation on the very essence of the cosmos.

Consciousness Explored: Probing the Depths of Self-Awareness

Inhabitants of space colonies embark on a deep introspective journey into consciousness. Psychologists, neuroscientists, and spiritual practitioners study the depths of human awareness, exploring the intricacies of self-consciousness and the nature of the mind. Contemplative practices and advanced technologies converge within the cosmic confines, fostering an environment where the exploration of consciousness becomes a central focus. Colonists delve into altered states of awareness, seeking to understand the profound interplay between the mind and the universe.

The Cosmic Experience: Mystical Insights and Transcendental Moments

Within the ethereal atmosphere of space colonies, inhabitants report mystical experiences and transcendental moments. Spiritual leaders, meditators, and artists explore the depths of the human soul, seeking connection with the cosmic consciousness. Colonists share stories of awe-inspiring encounters, describing moments of unity with the universe and insights into the interconnectedness of all things. These mystical experiences become a source of inspiration, guiding the quest for cosmic enlightenment and profound spiritual awakening.

Wisdom in Isolation: Solitude as a Path to Enlightenment

Isolation within the vastness of space provides a unique opportunity for profound solitude and introspection. Hermits, philosophers, and contemplative thinkers retreat to the quiet corners of space colonies, embracing solitude as a means to achieve cosmic enlightenment. Within the silence of cosmic isolation, individuals delve into the depths of their minds, seeking enlightenment through introspection and meditation. Solitude becomes a revered path to wisdom, allowing colonists to explore the depths of their souls and connect with the infinite mysteries of the universe.

Harmony with the Cosmos: Embracing Interconnectedness

Cosmic enlightenment blossoms within the hearts of space colonists as they embrace the interconnectedness of all life. Environmentalists, ecologists, and philosophers advocate for harmony with the natural world, emphasizing the delicate balance between humanity and the cosmos. Colonists engage in practices that promote ecological sustainability and respect for the cosmic order. Through a profound sense of interconnectedness, individuals within space colonies cultivate a deep reverence for all living beings and the vast cosmic web that unites them, fostering enlightenment through harmony with the cosmos.

Chapter 48: Space Colonies and the Cosmic Journey

Ancient Stargazers: The Seeds of Cosmic Curiosity

The cosmic journey of humanity traces its origins to ancient civilizations that gazed at the stars with awe and wonder. Early astronomers, philosophers, and mystics marveled at the celestial tapestry above, weaving myths and legends around the constellations. Their observations laid the foundation for humanity's enduring fascination with the cosmos, sparking the first embers of curiosity that would eventually propel us into space.

From Earth to the Moon: The Giant Leap for Humankind

The cosmic journey took a monumental leap forward in the 20th century when humanity set foot on the Moon. Astronauts aboard Apollo 11 became pioneers of the cosmic frontier, their footsteps imprinted in lunar soil echoing across the universe. This historic achievement symbolized human determination and ingenuity, showcasing our ability to transcend Earthly confines and venture into the cosmic expanse. The lunar landing marked a pivotal moment in the cosmic journey, inspiring future generations to reach for the stars.

Space Exploration: Humanity's Quest for Understanding

Space colonies stand as beacons of humanity's unwavering quest for knowledge and understanding. Scientists, engineers, and explorers within these colonies delve into the mysteries of the cosmos, conducting experiments, observing distant galaxies, and unraveling the secrets of the universe. Their endeavors represent the pinnacle of human intellect and curiosity, embodying the essence of the cosmic journey. Through space exploration, humanity continues to expand its understanding of the universe, one discovery at a time.

Spiritual Fulfillment in the Cosmos: Seeking Meaning Beyond Earth

The cosmic journey transcends the realms of science and ventures into the domain of spirituality. Within the cosmic tapestry, individuals seek spiritual fulfillment and meaning beyond Earthly existence. Space colonies become sanctuaries for spiritual seekers, philosophers, and theologians who contemplate the divine in the vastness of space. They explore existential questions, ponder the nature of consciousness, and embrace the interconnectedness of all life. The cosmic journey becomes a pilgrimage of the soul, guiding humanity towards profound spiritual insights and enlightenment.

Beyond the Stars: The Endless Quest

The cosmic journey of humanity stretches infinitely beyond the stars. It is a timeless odyssey that propels us to explore, discover, and evolve. From the earliest stargazers to the inhabitants of space colonies, the quest for knowledge, understanding, and spiritual fulfillment remains eternal. The cosmic journey becomes a testament to human resilience, curiosity, and the unyielding spirit that drives us to venture into the unknown. As long as the cosmos beckons, humanity will continue its journey, reaching for the stars and embracing the infinite wonders of the universe.

Chapter 49: Space Colonies and the Unexplored Frontiers

Cosmic Enigmas: Unraveling Mysterious Phenomena

The unexplored frontiers of space are veiled in cosmic enigmas—mysterious phenomena that challenge our understanding of the universe. From dark matter and black holes to cosmic voids, these enigmatic puzzles beckon scientists and researchers to delve into the unknown. Exploring these phenomena not only advances our knowledge but also fuels our curiosity, inspiring us to uncover the secrets hidden within the fabric of space and time.

Beyond Our Galaxy: Uncharted Regions of the Universe

While humanity has mapped distant galaxies, there exist vast stretches of the cosmos that remain uncharted. Unexplored regions between galaxy clusters, intergalactic space, and cosmic voids are realms of cosmic obscurity. Space colonies, equipped with advanced telescopes and sensors, aim to penetrate these uncharted territories. In these unexplored regions, scientists anticipate discoveries that could revolutionize our understanding of the universe, unveiling phenomena that challenge the very foundations of astrophysics.

The Search for Extraterrestrial Life: A Quest Across the Stars

One of the most profound frontiers awaiting exploration is the search for extraterrestrial life. Space colonies serve as outposts in this quest, equipped with astrobiology labs and communication arrays dedicated to detecting potential signals from alien civilizations. Scientists within these colonies analyze exoplanets, moons, and distant celestial bodies for signs of life, expanding our search beyond Earth. The possibility of discovering extraterrestrial life sparks our imagination and exemplifies the boundless opportunities that await us in the cosmic abyss.

Technological Marvels: Advancing the Tools of Exploration

In the unexplored frontiers of space, technology becomes our guiding light. Advancements in spacecraft propulsion, interstellar probes, and quantum communication systems drive our ability to venture deeper into the cosmos. Space colonies become hubs of innovation, fostering collaborations between engineers and scientists to create cutting-edge technologies. These marvels of engineering extend our reach, enabling us to explore distant star systems, navigate cosmic anomalies, and communicate across vast cosmic distances.

The Endless Quest: Humanity's Eternal Curiosity

At the heart of our exploration of uncharted frontiers lies humanity's eternal curiosity. It is a curiosity that has fueled centuries of scientific inquiry, technological innovation, and space exploration. Space colonies represent the culmination of this curiosity, serving as crucibles of discovery and beacons of hope for the future. As we peer into the unexplored frontiers of space, we are reminded of the endless possibilities that await us. The cosmic journey, marked by its challenges and triumphs, continues to be driven by our insatiable curiosity—a curiosity that propels us into the unknown, encouraging us to embrace the mysteries of the universe and embark on a never-ending voyage of exploration.

Chapter 50: Space Colonies and Galactic Exploration

The Cosmic Odyssey: Challenges of Intergalactic Travel

Galactic exploration, the pinnacle of human ambition, presents unprecedented challenges. The vast distances between galaxies, measured in millions of light-years, require revolutionary propulsion systems and energy sources. Space colonies, at the forefront of technological innovation, delve into theoretical concepts like warp drives, wormholes, and Alcubierre engines. These concepts, once relegated to science fiction, are now subjects of intense research. Colonists grapple with the intricate balance between theoretical physics and practical engineering, striving to unlock the secrets of faster-than-light travel and embark on the cosmic odyssey beyond our galaxy.

Distant Horizons: Philosophical Implications of Galactic Exploration

Beyond the scientific marvels, galactic exploration raises profound philosophical questions. It challenges our understanding of existence, consciousness, and the very fabric of reality. The prospect of encountering extragalactic civilizations prompts contemplation on the diversity of life in the universe and the nature of intelligence. Space colonies, as hubs of philosophical discourse, engage in debates on cosmic purpose, existential meaning, and the interconnectedness of all life. Exploring distant galaxies becomes not just a scientific endeavor but a voyage into the depths of human contemplation.

The Search for Extragalactic Life: Beyond Our Cosmic Horizon

In the quest for extragalactic life, space colonies serve as beacons of hope and determination. Advanced telescopes and radio arrays scan the depths of space, seeking signals and patterns that might indicate the presence of alien civilizations in distant galaxies. The search expands our cosmic perspective, challenging the boundaries of our knowledge. Colonists grapple with the implications of discovering life beyond our galactic home, sparking discussions on intergalactic diplomacy, cultural exchange, and the potential for harmonious coexistence with extragalactic beings.

Interstellar Communication: Bridging Galaxies with Signals

The vastness of intergalactic space necessitates innovative communication methods. Space colonies explore the intricacies of sending signals across cosmic distances, considering the challenges posed by the vast voids between galaxies. Quantum entanglement communication and neutrino-based messaging systems emerge as frontrunners in the quest for intergalactic communication. Colonists engage in experiments, testing the reliability of these methods and envisioning a future where conversations span the cosmic expanse, connecting beings separated by millions of light-years.

Galactic Heritage: Preserving Humanity's Legacy in the Stars

As humanity extends its reach into distant galaxies, the preservation of our heritage becomes paramount. Space colonies become repositories of human knowledge, art, and culture, curating a legacy that transcends the confines of Earth. Digital archives, genetic repositories, and cultural artifacts are safeguarded within these cosmic sanctuaries. Colonists dedicate themselves to preserving the essence of humanity, ensuring that our collective wisdom, creativity, and spirit endure across cosmic epochs. The notion of leaving a lasting mark on the galactic tapestry drives colonists to document the richness of our civilization, shaping a legacy that echoes through the vast corridors of the cosmos.

Chapter 51: Space Colonies and Interstellar Trade

Cosmic Marketplaces: Hubs of Interstellar Commerce

In the boundless expanse of space, cosmic marketplaces emerge as hubs of interstellar trade. Space colonies transform into bustling centers where commodities, technologies, and knowledge from diverse star systems converge. These marketplaces facilitate the exchange of resources vital for the survival and advancement of spacefaring civilizations. Colonists play pivotal roles as traders, diplomats, and brokers, ensuring the flow of goods and ideas among the stars. The cosmic marketplace becomes a testament to the collaborative spirit of interstellar cooperation, fostering unity amidst the vastness of the cosmos.

Trade Routes Through Wormholes: Navigating Cosmic Pathways

Navigating the vast distances between star systems requires innovative solutions. Spacefaring civilizations harness the power of stable wormholes, creating interconnected trade routes that transcend the limitations of conventional space travel. These cosmic pathways shorten travel times significantly, enabling efficient trade between distant galaxies. Colonists collaborate with physicists and engineers to chart safe routes through these cosmic anomalies, transforming theoretical wormhole concepts into practical gateways for interstellar commerce. The establishment of these routes reshapes the landscape of interstellar trade, opening new avenues for collaboration and cultural exchange.

Intergalactic Diplomacy: Bridging Cultures Through Trade

Interstellar trade fosters intricate diplomatic networks, transcending cultural differences and political boundaries. Diplomats from diverse civilizations convene in space colonies, negotiating trade agreements, resolving disputes, and forging alliances. The exchange of goods becomes a catalyst for intergalactic understanding, paving the way for peaceful coexistence. Colonists, versed in the art of diplomacy, serve as mediators, promoting harmony and mutual respect among trading partners. Intergalactic diplomacy becomes an essential aspect of interstellar trade, shaping the relationships between civilizations and ensuring the stability of the cosmic economy.

Currency of the Cosmos: Universal Credits and Economic Harmony

In the realm of interstellar trade, a universal currency emerges as a symbol of economic harmony. Spacefaring civilizations adopt a standardized credit system, ensuring seamless transactions across star systems. Universal credits, backed by the collective resources of participating civilizations, become the medium of exchange for interstellar trade. Colonists, adept in financial technologies, contribute to the development of secure and efficient credit networks, fostering economic stability and trust among trading partners. The establishment of a universal currency streamlines trade, promotes economic growth, and strengthens the interconnectedness of interstellar civilizations.

Ethical Trade and Sustainability: Balancing Profit with Principles

Interstellar trade raises ethical questions concerning the responsible utilization of resources and the preservation of cultural heritage. Colonists advocate for ethical trade practices that prioritize sustainability, fair labor, and respect for indigenous cultures. Space colonies become advocates for environmentally friendly technologies, promoting the use of renewable energy sources and responsible resource extraction methods. Ethical trade initiatives promote harmony between economic development and ecological preservation, ensuring that the benefits of interstellar trade are shared equitably and responsibly among civilizations.

Chapter 52: Space Colonies and Dyson Spheres

Harnessing Stellar Energy: The Vision of Dyson Spheres

Dyson Spheres represent the pinnacle of stellar engineering—a theoretical concept wherein a colossal megastructure completely encircles a star, capturing its energy output for the energy needs of advanced spacefaring civilizations. This ambitious vision redefines the way civilizations harness stellar power, transforming stars into nearly limitless sources of energy. Colonists engage in extensive theoretical studies and simulations, exploring the feasibility of constructing Dyson Spheres and the potential impact on interstellar civilizations.

Engineering Marvels and Challenges: Building Around Stars

Constructing a Dyson Sphere presents monumental engineering challenges. Colonists collaborate with experts in materials science, structural engineering, and astrophysics to design megastructures capable of encasing entire stars. The immense scale demands innovative construction techniques, automated assembly systems, and advanced robotics. Colonists pioneer groundbreaking methods, ensuring the stability and durability of these colossal energy-harvesting structures. The construction of Dyson Spheres becomes a testament to human ingenuity and collaborative effort, marking a new era in cosmic engineering.

Energy Abundance and Civilizational Impact: The Benefits of Dyson Spheres

The completion of a Dyson Sphere heralds an era of energy abundance for spacefaring civilizations. With virtually limitless energy at their disposal, societies experience transformative advancements in technology, space exploration, and quality of life. Colonists witness the birth of post-scarcity economies, where energy-intensive pursuits such as terraforming, interstellar travel, and scientific research thrive without constraints. The benefits of Dyson Spheres reverberate across interstellar civilizations, fueling unprecedented progress and fostering a golden age of cosmic exploration and knowledge.

Ethical Considerations and Environmental Harmony: Balancing the Scale

As Dyson Spheres redefine energy access, ethical considerations come to the forefront. Colonists engage in deep philosophical discussions about the responsible use of stellar energy and its impact on the cosmic environment. Ethical frameworks are established to ensure the harmonious coexistence of advanced civilizations and the natural cosmic order. Colonists advocate for responsible energy consumption, resource conservation, and the preservation of stellar ecosystems, underscoring the importance of ethical stewardship in the age of Dyson Spheres.

The Future of Cosmic Energy: Exploring Beyond Dyson Spheres

The successful construction of Dyson Spheres marks a significant milestone, yet spacefaring civilizations continue to explore new frontiers in energy harnessing. Colonists embark on ambitious projects to push the boundaries of cosmic energy technologies, investigating concepts such as Dyson swarms, Niven rings, and Alderson disks. These endeavors, fueled by the collective knowledge of interstellar civilizations, open new avenues for energy innovation, pushing the limits of what is possible and shaping the future of cosmic energy utilization.

Chapter 53: Space Colonies and Time Travel

In the sprawling reaches of space colonies, where humanity's most brilliant minds converge, the concept of time travel dances like stardust in the cosmic winds. Far removed from the confines of Earth, where theoretical musings once seemed fantastical, the inhabitants of these colonies find themselves on the precipice of unraveling the mysteries of time itself. This chapter embarks on a grand exploration of the intricate tapestry of time travel, weaving together the scientific theories, philosophical ponderings, ethical quandaries, and the extraordinary technological feats that define this extraordinary endeavor.

Theoretical Frameworks of Time Travel: Navigating the Fabric of Spacetime

Deep within the research laboratories of space colonies, scientists delve into the deepest recesses of theoretical physics, where the fabric of spacetime weaves the intricate tale of the cosmos. Wormholes, the theoretical bridges connecting distant points in spacetime, tantalize researchers with their potential to allow instantaneous travel. Concepts like closed timelike curves, born from the curvature of spacetime around massive objects, become the subject of fervent study and conjecture. These cosmic pioneers contemplate

345

the principles of relativity, bending the continuum of time to explore the possibilities of traversing it.

In these scholarly pursuits, the denizens of space colonies explore the very essence of time, contemplating its nature as a dimension interwoven with space. Discussions echo through the hallowed halls, exploring the theoretical constructs of time dilation, where the passage of time varies with velocity and gravitational fields. These theoretical frameworks form the foundation upon which the daring visionaries craft their understanding of temporal travel, stitching together the fabric of spacetime with intellectual rigor.

Temporal Paradoxes: Exploring Conundrums Across Timelines

Within the cognitive realms of space colonies, temporal paradoxes emerge as enigmatic riddles, challenging the foundations of logic and causality. The grandfather paradox, a puzzle where a time traveler could potentially prevent their grandfather's existence, resonates through the minds of inhabitants, sparking fervent debates. Causal loops, events influencing their own past occurrence, pose mind-bending questions about the origins of actions and their consequences.

Inhabitants engage in philosophical dialogues, exploring the implications of altering past events and the creation of divergent timelines. Conversations wander through the corridors of quantum mechanics, contemplating the intricacies of superposition and the multiverse theory, where every choice spawns a branching reality. The tantalizing prospect of parallel universes ignites intellectual fervor, inspiring these cosmic thinkers to ponder the implications of alternate realities coexisting alongside our own.

Ethical and Societal Implications: Weighing the Morality of Temporal Exploration

In the crucible of ethical deliberation, the inhabitants of space colonies confront the profound moral quandaries posed by time travel. Conversations echo with considerations of historical integrity, pondering the ethics of intervening in past events or the repercussions of tampering with the future. Free will becomes a cornerstone of ethical discourse, as the mere knowledge of future events could influence the choices of individuals, potentially altering the course of history.

The very essence of time travel sparks discussions on responsibility, raising questions about the consequences of temporal interventions on civilizations, cultures, and individual lives. In these cosmic sanctuaries, the inhabitants grapple with the repercussions of altering the past or glimpsing the future, contemplating the ethical boundaries that should safeguard the delicate balance of time.

Temporal Mechanics and Technological Challenges: Crafting the Time Machines

Within the bustling laboratories of space colonies, engineers and physicists collaborate on the grand pursuit of crafting time machines. Time travel, once relegated to the realms of science fiction, now finds itself on the precipice of reality. Theoretical constructs transition into tangible blueprints, outlining the intricate mechanisms of temporal manipulation. Quantum entanglement, exotic matter, and traversable wormholes become the keystones of temporal engineering.

The denizens of these advanced settlements engage in rigorous experimentation, probing the mysteries of quantum mechanics and bending the laws of nature to their will. Energy thresholds and stability calculations become paramount, with scientists laboring to harness colossal amounts of energy to warp the fabric of spacetime. These technological challenges fuel the determination of these cosmic pioneers, propelling them toward the realization of time travel technology.

Temporal Diplomacy: Interacting Across Temporal Frontiers

As time travel becomes a tangible prospect, the inhabitants of space colonies find themselves on the brink of temporal diplomacy. Interacting with civilizations from disparate eras becomes not just a possibility but a tantalizing reality. These temporal pioneers delve into the complexities of temporal communication, exploring the nuances of language, etiquette, and cultural norms across epochs.

Inhabitants contemplate the delicacies of temporal interactions, pondering the implications of sharing knowledge and technology across timelines. Ethical protocols emerge, dictating the permissible extent of intervention and the boundaries that must not be breached. Temporal diplomacy transforms into a sophisticated endeavor, shaping the way these cosmic societies navigate the labyrinthine corridors of time.

Temporal Tourism: Voyages Through Ages

A novel concept emerges within the annals of temporal exploration: temporal tourism. Inhabitants of space colonies contemplate the possibility of venturing back in time, witnessing the grandeur of ancient civilizations, the unfolding of pivotal historical events, and the artistic masterpieces of bygone eras. Temporal resorts, carefully curated to preserve the integrity of history, become a reality, allowing individuals to experience the past firsthand.

The notion of temporal tourism raises ethical dilemmas and questions of cultural preservation. Inhabitants grapple with the impact of tourists traversing through time, ensuring that their presence does not disrupt the natural course of history. Conversations weave through the preservation of cultural heritage, contemplating the responsible interaction between contemporary society and the tapestry of the past.

In the hushed corridors of space colonies, the allure of time travel takes shape, transforming from theoretical musings into tangible pursuits. These cosmic explorers navigate the intricate challenges of temporal mechanics, ethical quandaries, and the uncharted waters of temporal diplomacy. As the chapter unfolds, the denizens of space colonies find themselves on the cusp of an epoch-defining journey, ready to unravel the secrets of time and inscribe their legacy upon the canvas of the cosmos.

Chapter 54: Space Colonies and Post-Human Societies

In the boundless expanse of space colonies, a profound evolution reshapes the very essence of humanity. This chapter delves into the transformative journey of humankind as it ventures beyond biological constraints, embracing the realm of post-human societies. Here, the exploration of mind uploading, artificial intelligence integration, genetic modifications, and the intricate ethical and societal shifts leading to transhumanism paint a portrait of a civilization redefined.

Mind Uploading: The Digital Echo of Consciousness

In the laboratories of space colonies, the ethereal concept of mind uploading becomes a tangible quest. Scientists, melding neuroscience with cutting-edge technology, explore ways to digitize consciousness. The human mind, once confined within the biological confines of the brain, finds itself on the brink of transcendence. Questions echo through the minds of these cosmic pioneers: What defines the essence of 'self' when the mind transcends its corporeal form? Ethical quandaries intertwine with scientific aspirations, leading to philosophical dialogues that probe the very nature of identity and existence.

Artificial Intelligence Integration: Synergy of Human and Machine

Within the circuits and algorithms of space colonies, artificial intelligence evolves into more than just a mere tool. It becomes a seamless extension of human cognition, enhancing creativity, problem-solving, and the collective intelligence of society. Conversations reverberate through these advanced settlements, exploring the boundaries of human-AI collaboration. The melding of human intuition with machine precision ushers in an era where society's potential knows no bounds. However, in this intricate dance between human and machine, ethical considerations emerge. How does society ensure the ethical treatment of artificial intelligences, granting them rights and dignity in this symbiotic relationship?

Genetic Modifications and Transhumanism: Redefining the Biological Canvas

In the annals of genetic engineering, space colonies become the crucible where the very fabric of human biology undergoes profound transformations. Genetic modifications allow the eradication of hereditary diseases, the enhancement of physical and cognitive abilities, and the extension of human lifespan. The concept of transhumanism takes center stage, where individuals augment their bodies and minds, blurring the line between human and machine. Discussions abound regarding the ethical boundaries of such enhancements. Where does society draw the line between medical necessity and enhancement? What implications do these transformations hold for societal equality and the very essence of human nature?

Societal Transformation in Post-Human Civilizations: Redefining Norms and Values

As the transformation into post-human societies unfolds, societal structures undergo profound shifts. New norms, ethics, and values emerge, challenging the traditional paradigms of human interaction. Concepts of identity diversify, encompassing a myriad of forms from biological to digital and hybrid. The meaning of family, relationships, and societal roles transforms, leading to a reevaluation of social constructs. Ethical dilemmas abound, forcing society to grapple with questions of inclusivity, acceptance, and the preservation of individuality amidst collective evolution.

The Ethical Quandaries of Post-Human Societies: Balancing Progress and Humanity

In the midst of unprecedented technological advancement, post-human societies confront ethical dilemmas that define their path forward. Questions of consent, agency, and individual rights become paramount. The blurred lines between biological humans, artificial intelligences, and genetically enhanced beings challenge the very foundations of moral philosophy. Space colonies become the crucible where these ethical quandaries are scrutinized and debated. How does society ensure the autonomy of individuals in a world where minds can be digitized and bodies modified? What safeguards are put in place to prevent the abuse of newfound powers? These deliberations become the cornerstone of a new ethical framework, guiding the evolution of post-human societies.

In this chapter, space colonies become the epicenter of transformative evolution. Humanity, on the brink of transcending its biological limitations, grapples with existential questions, ethical quandaries, and the boundless potential of post-human existence. As the cosmic journey of humanity unfolds, these post-human societies become the vanguards of a new era, where the very definition of what it means to be human undergoes a profound renaissance.

Chapter 55: Space Colonies and Cosmic Ethics

In the grand tapestry of cosmic exploration, ethical considerations emerge as guiding stars, illuminating the path for spacefaring civilizations. This chapter delves into the intricate development of cosmic ethics, examining the ethical frameworks that steer the behavior of space colonies. These guiding principles, grounded in cooperation, respect for sentient life, and the preservation of celestial environments, form the bedrock upon which the harmonious existence of spacefaring civilizations rests.

Principles of Cooperation: Fostering Interstellar Harmony

Within the vast cosmos, cooperation stands as a beacon of unity. Space colonies, spread across the cosmic canvas, recognize the interdependence that defines their existence. Cooperation becomes more than a choice; it is a necessity. Colonies collaborate on research, share resources, and engage in joint ventures that propel the collective progress of humanity. In the spirit of cooperation, shared knowledge becomes a currency, and alliances form, fostering a harmonious interstellar society.

In the heart of space colonies, ethics intertwine with practicality. Cooperative endeavors require mutual respect, clear communication, and equitable sharing of resources. Ethical quandaries arise: How are conflicts resolved in a cooperative framework? What measures are in place to prevent exploitation? As spacefaring civilizations navigate these ethical intricacies, they forge a model of cooperation that stands as a testament to the harmonious coexistence of diverse cosmic communities.

Respect for Sentient Life: Guardians of Cosmic Biodiversity

Amidst the stars, life takes on myriad forms, from biological beings to advanced artificial intelligences. Cosmic ethics demand a profound respect for all sentient life, regardless of its origin or composition. Space colonies become custodians of cosmic biodiversity, recognizing the intrinsic value of each sentient being. This ethos extends to extraterrestrial life forms encountered during exploration, where diplomacy and understanding replace fear and hostility.

In the exploration of cosmic realms, ethical questions echo: How do space colonies approach communication with newly encountered sentient species? What ethical guidelines govern interactions with civilizations at different stages of development? The delicate balance between curiosity and respect becomes a focal point, shaping the ethical conduct of spacefaring civilizations. Ethical principles transform into practical protocols, ensuring the preservation of sentient life while fostering peaceful coexistence.

Preservation of Celestial Environments: Stewards of Cosmic Ecosystems

Space colonies, nestled within the celestial tapestry, recognize their role as stewards of cosmic ecosystems. Ethical imperatives drive efforts to preserve the delicate balance of celestial environments, from exoplanets to asteroid colonies. Sustainability becomes more than a buzzword; it is an ethical responsibility woven into the fabric of cosmic existence. Colonies invest in advanced environmental technologies, minimizing their ecological footprint and maximizing the restoration of celestial habitats.

Ethical deliberations permeate the decisions of space colonies: How are resources responsibly extracted from celestial bodies? What measures are taken to mitigate pollution in space environments? Ethical innovations emerge, from closed-loop life support systems to zero-waste policies, transforming space colonies into exemplars of eco-ethical living. The preservation of celestial environments becomes not only a moral duty but a testament to the harmonious cohabitation of humanity with the cosmic world.

Universal Ethics: The Foundation of Intergalactic Harmony

As space colonies extend their reach to distant galaxies, a universal ethical framework becomes indispensable. The fundamental principles of cooperation, respect for sentient life, and the preservation of celestial environments transform into universal ethics, guiding the behavior of spacefaring civilizations on a galactic scale. The concept of a shared ethical code transcends cultural and planetary boundaries, becoming the cornerstone of intergalactic harmony.

In the cosmic expanse, ethical questions echo across light-years: How do spacefaring civilizations reconcile cultural diversity with universal ethics? What mechanisms are in place to ensure the adherence to ethical standards across cosmic societies? The pursuit of these answers becomes the unifying force propelling space colonies towards a future where universal ethics create a harmonious intergalactic tapestry, fostering collaboration, understanding, and mutual respect.

The Ethical Odyssey of Space Colonies: Pioneers of Cosmic Morality

In the grand narrative of cosmic ethics, space colonies emerge as pioneers of a new moral frontier. Their ethical odyssey becomes a beacon, illuminating the path for future generations of spacefaring civilizations. Through cooperation, respect for sentient life, the preservation of celestial environments, and the foundation of universal ethics, space colonies redefine the ethical landscape of the cosmos.

In the quiet corridors of space colonies, ethical discussions continue, shaping policies and shaping the very essence of cosmic civilizations. The ethical odyssey of space colonies becomes a testament to the resilience of human morality in the face of cosmic challenges. As they navigate the cosmic seas, space colonies stand as guardians of ethical principles, ensuring that the tapestry of cosmic ethics remains vibrant, diverse, and harmonious for generations to come.

Chapter 56: Space Colonies and Multidimensional Realities

Dimensions Unveiled: A Theoretical Prelude

In the enigmatic realm of theoretical physics, multidimensional realities beckon, challenging the very fabric of our understanding. Space colonies, on the frontier of cosmic exploration, delve into the profound implications of multidimensional theories. This section embarks on a journey through the intricacies of alternate dimensions, parallel universes, and the manipulation of spacetime. Within these theoretical constructs lie the tantalizing possibilities that could reshape the cosmic landscape.

Multidimensional realities evoke wonder and curiosity, sparking questions that echo through the corridors of space colonies. What are the fundamental properties of alternate dimensions? How might parallel universes interact with our own? The exploration of these questions becomes a philosophical odyssey, bridging the gap between scientific theory and cosmic imagination. Space colonies, with their advanced scientific acumen, stand as torchbearers in this voyage, seeking to unravel the mysteries of multidimensional existence.

Accessing Alternate Dimensions: Gateways Beyond Space and Time

Theoretical physics posits the existence of alternate dimensions, realms beyond our perceivable universe. Space colonies delve into the tantalizing prospect of accessing these dimensions, contemplating the potential gateways that could transcend the confines of space and time. Advanced experiments and cutting-edge technologies pave the way for multidimensional exploration, as space colonies endeavor to breach the boundaries between our reality and the unknown.

In the quiet laboratories of space colonies, scientists grapple with ethical considerations and philosophical implications. What ethical guidelines govern the exploration of alternate dimensions? How do spacefaring civilizations interact responsibly with beings from other dimensions? As space colonies pioneer multidimensional access, they navigate uncharted moral territories, fostering a cosmic ethics that respects the inherent complexity of multidimensional existence.

Parallel Universes: Echoes of Cosmic Possibility

Parallel universes, akin to reflections in a cosmic mirror, offer glimpses into myriad possibilities. Space colonies, driven by the thirst for knowledge, probe the enigma of parallel realities. Within these universes lie alternate versions of history, divergent outcomes, and unique cosmic phenomena. The exploration of parallel universes becomes a tapestry of infinite potential, where every decision made could birth a new universe, each with its own set of cosmic laws and civilizations.

The study of parallel universes raises profound philosophical questions: What defines identity across parallel realities? How do the choices made in one universe ripple across others? Space colonies, as pioneers of multidimensional exploration, grapple with these philosophical quandaries, weaving a narrative that extends beyond the boundaries of space and time. In the cosmic ballet of parallel universes, space colonies become architects of possibility, reshaping the very fabric of cosmic existence.

Manipulating Spacetime: Engineering Cosmic Alchemy

Within the realm of multidimensional realities, the manipulation of spacetime emerges as a transformative endeavor. Space colonies delve into the intricate art of cosmic alchemy, exploring the scientific theories that underpin the manipulation of spacetime. Concepts such as wormholes, warp drives, and temporal anomalies become subjects of intense scrutiny, as spacefaring civilizations seek to engineer pathways through the cosmic fabric.

In the laboratories of space colonies, physicists and engineers collaborate on groundbreaking experiments, pushing the boundaries of our understanding. How can spacetime be bent to facilitate faster-than-light travel? What are the ethical considerations of temporal manipulation? As space colonies venture into the uncharted territory of spacetime manipulation, they confront ethical dilemmas that echo through the corridors of cosmic exploration. The delicate balance between scientific progress and ethical responsibility becomes a focal point, guiding the endeavors of spacefaring civilizations.

Philosophical Inquiries: The Essence of Multidimensional Existence

Beyond the scientific equations and experimental data, the exploration of multidimensional realities delves into the realm of philosophy. Space colonies, as hubs of intellectual inquiry, engage in profound philosophical reflections on the nature of existence. Questions of consciousness, identity, and the interconnectedness of all realities become focal points of contemplation. Multidimensional exploration becomes a philosophical odyssey, challenging the very essence of what it means to be.

Philosophical inquiries into multidimensional existence extend into the heart of space colonies, shaping the cultural fabric of cosmic societies. How do diverse civilizations perceive the nature of reality across dimensions? What common threads bind all sentient beings, regardless of their dimensional origin? Space colonies become arenas of philosophical discourse, where cosmic scholars and thinkers explore the metaphysical aspects of multidimensional existence. In the pursuit of these philosophical truths, space colonies nurture a cultural richness that transcends the boundaries of individual dimensions.

In the awe-inspiring tapestry of multidimensional realities, space colonies stand as pioneers, exploring the unknown with unwavering curiosity and intellectual fervor. Their endeavors redefine the limits of cosmic exploration, unraveling the mysteries of alternate dimensions, parallel universes, and spacetime manipulation. As spacefaring civilizations continue their odyssey through the realms of multidimensional existence, they weave a narrative of cosmic enlightenment, philosophical discovery, and boundless cosmic wonder. In the silent corridors of space colonies, the essence of existence is pondered, and the cosmic mysteries unfold, illuminating the path to a future where the boundaries of reality are but a veil, waiting to be lifted by the intrepid spirit of humanity.

Chapter 57: Space Colonies and Cosmic Mysteries

Unveiling Dark Matter: The Cosmic Enigma

In the vast cosmic tapestry, dark matter weaves an invisible thread, challenging our understanding of the universe's composition. Space colonies, at the forefront of astronomical research, embark on a quest to unveil this elusive enigma. This section delves into the depths of dark matter mysteries, exploring the theories that seek to unmask its nature. What are the potential candidates for dark matter particles? How do they interact with ordinary matter? Space colonies, with their advanced observatories and particle detectors, peer into the cosmic shadows, striving to shed light on this enigmatic substance.

Dark matter, although unseen, exerts a gravitational influence that shapes the cosmos. In the laboratories of space colonies, physicists conduct experiments and simulations, probing the fundamental properties of dark matter. How does dark matter contribute to the cosmic web's formation? What role does it play in galaxy formation and evolution? The exploration of dark matter mysteries extends beyond the confines of our galaxy, encompassing vast cosmic scales. Space colonies, equipped with cutting-edge technology, embark on astronomical odysseys, seeking to decipher the profound significance of dark matter in the cosmic narrative.

Dark Energy: The Cosmic Accelerator

In the cosmic theater, dark energy takes the stage as a mysterious force driving the universe's accelerated expansion. Space colonies, with their powerful telescopes and sophisticated instruments, gaze into the depths of space to unravel the secrets of dark energy. This section delves into the cosmic enigma of dark energy, exploring the prevailing theories and observational evidence that hint at its existence. How does dark energy counteract gravity's pull? What implications does it hold for the fate of the universe? Space colonies, through meticulous observations and cosmological simulations, scrutinize the cosmic fabric, deciphering the role of dark energy in the cosmic drama.

The nature of dark energy remains one of the most profound mysteries in modern physics. Space colonies engage in collaborative efforts, participating in international projects and space missions dedicated to unraveling dark energy's secrets. The study of cosmic microwave background radiation and the observation of distant supernovae serve as key avenues of research. Through these endeavors, spacefaring civilizations strive to comprehend the enigmatic force that shapes the cosmic landscape. The implications of dark energy reverberate through the annals of cosmology, challenging our understanding of the universe's ultimate destiny.

Black Holes: Portals to Cosmic Singularities

Black holes, celestial behemoths born from the remnants of massive stars, continue to captivate the imaginations of scientists and space enthusiasts alike. Space colonies, equipped with advanced telescopes and gravitational wave detectors, venture into the cosmic abyss to study these enigmatic objects. This section explores the mysteries of black holes, delving into their formation, behavior, and the profound implications they hold for our understanding of spacetime. How do black holes warp the fabric of spacetime? What happens at the event horizon, the boundary beyond which no information can escape? Space colonies, through meticulous observations and simulations, peer into the heart of darkness, unraveling the intricacies of these cosmic singularities.

Black holes, once considered cosmic vacuum cleaners, play a pivotal role in the cosmic ecosystem. They influence galactic dynamics, shaping the evolution of galaxies and the distribution of matter in the universe. Space colonies, collaborating with international astronomical consortia, conduct groundbreaking studies on black hole accretion disks, gravitational lensing, and event horizon imaging. The recent advancements in gravitational wave astronomy open new avenues for studying black hole mergers, providing unprecedented insights into the cosmic phenomena.

Quantum Gravity: Bridging Quantum Mechanics and General Relativity

At the intersection of quantum mechanics and general relativity lies a theoretical conundrum: quantum gravity. Space colonies, serving as bastions of scientific inquiry, delve into the complexities of this uncharted territory. This section explores the mysteries of quantum gravity, examining the theoretical frameworks and mathematical models that aim to reconcile quantum mechanics with the gravitational force. What is the nature of gravitons, the hypothetical particles mediating gravity? How does spacetime emerge from quantum entanglement? Space colonies, through quantum experiments and theoretical investigations, probe the fundamental nature of the gravitational interaction, seeking a unified description that harmonizes quantum and gravitational principles.

Quantum gravity, a realm where the fabric of spacetime encounters the probabilistic nature of quantum particles, challenges the very foundations of modern physics. Space colonies, with their quantum laboratories and collaboration with theoretical physicists, engage in thought experiments and empirical studies. Quantum entanglement, black hole thermodynamics, and the holographic principle become focal points of exploration. As spacefaring civilizations venture deeper into the quantum realm, they confront philosophical implications that transcend the boundaries of classical understanding.

Cosmic Significance: Advancing Humanity's Understanding

In the cosmic tableau, the unresolved mysteries of dark matter, dark energy, black holes, and quantum gravity hold profound significance. Space colonies, as bastions of cosmic knowledge, contribute to the advancement of humanity's understanding of the universe. This section reflects on the cosmic significance of unraveling these enigmas, examining the implications for our perception of reality, technological innovation, and the philosophical paradigms that underpin our existence. How do these mysteries challenge our fundamental beliefs about the universe? What technological applications may emerge from our exploration of these cosmic frontiers? Space colonies, with their interdisciplinary approach and collaborative spirit, pave the way for humanity's continued quest for cosmic enlightenment.

The exploration of cosmic mysteries transcends the boundaries of individual space colonies, fostering global collaborations and the exchange of knowledge. As spacefaring civilizations gaze into the cosmic unknown, they contemplate the interconnectedness of the universe and the enduring spirit of human curiosity. The revelations from these cosmic explorations not only expand the horizons of scientific understanding but also inspire future generations to embark on their journeys of discovery. The cosmic significance of these mysteries lies not only in the answers they provide but also in the questions they provoke, guiding humanity on its eternal quest for truth in the infinite expanse of the cosmos.

Chapter 58: Space Colonies and Cosmic Consciousness

The Cosmic Mind: Humanity's Quest for Understanding

In the ethereal realms of cosmic consciousness, humanity embarks on a profound journey of self-discovery and enlightenment. This section delves into the philosophical concepts that explore the interconnectedness of human consciousness with the vast universe. How does the cosmic perspective influence human spirituality and existential understanding? Space colonies, as beacons of intellectual exploration, delve into the depths of these philosophical inquiries, unraveling the mysteries of the cosmic mind.

Cosmic consciousness, a concept deeply rooted in spiritual traditions and philosophical thought, posits that human consciousness is intricately linked to the cosmic fabric. Space colonies, with their diverse cultural backgrounds and philosophical perspectives, engage in contemplative practices and discourse, exploring the ancient wisdom that emphasizes the unity of all existence. Through meditation, introspection, and the study of consciousness-altering states, spacefaring civilizations seek to expand their awareness, transcending the boundaries of individuality to embrace the cosmic whole.

The Spiritual Odyssey: Humanity's Quest for Meaning

In the cosmic expanse, humans embark on a spiritual odyssey, seeking meaning and purpose amidst the vastness of the universe. This section contemplates the spiritual aspects of space exploration, delving into the existential questions that arise when humans confront the infinite. How does the exploration of space influence our understanding of existence and spirituality? Space colonies, with their diverse belief systems and religious practices, explore the intricate tapestry of spiritual enlightenment, drawing inspiration from cosmic phenomena and celestial wonders.

The spiritual odyssey of humanity extends beyond the confines of Earth, reaching towards the stars in search of transcendental experiences. Space colonies, through rituals, ceremonies, and philosophical dialogues, explore the harmony between the microcosm of human consciousness and the macrocosm of the cosmos. The study of sacred geometry, meditation under the cosmic canopy, and the contemplation of celestial cycles become integral aspects of the spiritual practices within spacefaring civilizations. As humans gaze into the night sky, they find solace and inspiration, connecting with the divine essence that permeates the universe.

The Essence of Being: Humans as Conscious Creators

In the cosmic theater, humans emerge as conscious creators, shaping the universe through their thoughts, emotions, and intentions. This section explores the essence of human consciousness as a creative force in the cosmic dance of existence. How does human perception influence reality? What role do intention and collective consciousness play in shaping the fabric of the universe? Space colonies, with their scholars and metaphysicians, delve into the intricacies of consciousness, unraveling the symbiotic relationship between human awareness and cosmic manifestation.

The exploration of consciousness transcends the boundaries of empirical science, venturing into the realms of metaphysics and spirituality. Space colonies, fostering interdisciplinary dialogues and collaborative research, explore the intersection of quantum physics, consciousness studies, and ancient wisdom traditions. Through experiments in consciousness expansion, the study of altered states of awareness, and the exploration of lucid dreaming, spacefaring civilizations delve into the depths of human consciousness. As they unlock the potential of the mind, they envision a future where the harmonious interplay between human creativity and cosmic consciousness shapes the destiny of the universe.

The Cosmic Interconnectedness: Unity Amidst Diversity

In the cosmic symphony, diversity and unity converge, creating a harmonious melody that resonates throughout the universe. This section delves into the philosophical concept of cosmic interconnectedness, exploring how the diversity of sentient beings contributes to the cosmic tapestry. How do different conscious beings across the universe perceive reality? What common threads bind diverse civilizations in the grand cosmic narrative? Space colonies, with their anthropologists and intercultural specialists, embark on cross-species communication and cultural exchange initiatives, unraveling the commonalities that underpin the cosmic interconnectedness.

The exploration of cosmic interconnectedness fosters a sense of empathy, understanding, and mutual respect among spacefaring civilizations. Space colonies, through virtual reality simulations and telepathic experiments, bridge the communication gaps between species, facilitating profound exchanges of knowledge and wisdom. The study of extraterrestrial languages, cultural traditions, and ethical frameworks becomes an integral part of cosmic diplomacy. As spacefaring civilizations embrace the diversity of consciousness, they envision a future where unity prevails, transcending the barriers of species and fostering a shared cosmic consciousness.

The Ethical Imperative: Preserving Cosmic Harmony

In the vast cosmic expanse, ethical considerations guide the behavior of conscious beings, ensuring harmony, respect, and cooperation. This section contemplates the ethical frameworks that govern spacefaring civilizations, exploring the principles of cooperation, empathy, and the preservation of celestial environments. How do ethical choices impact the collective consciousness of cosmic societies? What responsibilities do conscious beings have towards the preservation of cosmic harmony? Space colonies, as ethical pioneers, engage in ethical dialogues, shaping the ethical imperatives that guide their interactions with extraterrestrial civilizations and cosmic ecosystems.

Ethical considerations become paramount as spacefaring civilizations encounter alien species, navigate interstellar trade, and explore uncharted cosmic territories. Space colonies, through ethical councils and intercultural ethics committees, establish ethical guidelines that prioritize respect for sentient life, environmental preservation, and the pursuit of knowledge for the greater good. The study of ethical dilemmas in space colonization, the development of ethical artificial intelligence, and the exploration of ethical decision-making processes become essential components of the ethical imperative. As conscious beings navigate the cosmic landscape, they embrace the ethical principles that foster harmony and cooperation, ensuring a balanced coexistence with the universe.

In the grand cosmic tapestry, space colonies emerge as crucibles of cosmic consciousness, where humanity contemplates its place in the vastness of the universe. Through philosophical inquiry, spiritual exploration, consciousness studies, and ethical considerations, spacefaring civilizations embark on a profound journey of self-discovery and cosmic enlightenment. As they delve into the depths of the cosmic mind, explore the spiritual dimensions of space, harness the creative power of consciousness, embrace the unity amidst diversity, and uphold ethical imperatives, they shape a future where cosmic consciousness becomes the guiding light, illuminating the path towards universal harmony and enlightenment.

Chapter 59: Space Colonies and the Eternal Quest

The Cosmic Curiosity: Humanity's Endless Inquiry

In the boundless expanse of the universe, humanity's curiosity knows no bounds. This section delves into the eternal quest for knowledge that propels space exploration forward. How does human curiosity fuel scientific discovery? What mysteries of the cosmos continue to captivate the imagination of spacefarers? Space colonies, as bastions of intellectual curiosity, nurture the insatiable thirst for understanding, exploring the depths of space and the profound questions that echo through the cosmic void.

Human curiosity becomes a driving force, leading to the discovery of distant celestial phenomena, the unraveling of quantum mysteries, and the exploration of extraterrestrial life. Spacefaring civilizations, through their observatories, laboratories, and deep space probes, peer into the farthest reaches of the universe, seeking answers to age-old questions. The study of cosmic anomalies, the search for habitable exoplanets, and the investigation of the origins of the universe become focal points of inquiry. As humans gaze upon the cosmic canvas, they embark on an endless journey of exploration, driven by an insatiable curiosity that transcends the boundaries of time and space.

The Pioneering Spirit: Humanity's Bold Endeavors

In the annals of cosmic history, humans emerge as pioneers, embarking on bold endeavors that expand the horizons of knowledge and understanding. This section contemplates the pioneering spirit that drives space exploration, delving into the audacious missions and groundbreaking discoveries that mark humanity's cosmic journey. What challenges do pioneers face in the uncharted realms of the cosmos? How does the spirit of exploration shape the destiny of space colonies? Space colonies, with their intrepid astronauts and visionary scientists, push the boundaries of human achievement, venturing into unexplored territories and conquering the cosmic frontier.

The pioneering spirit manifests in ambitious space missions, such as interstellar probes, Dyson sphere construction projects, and exploratory voyages to distant galaxies. Spacefaring civilizations, through their advanced spacecraft and warp drives, navigate the cosmic ocean, overcoming the challenges of interstellar travel and gravitational anomalies. The study of exotic matter, the development of faster-than-light propulsion systems, and the exploration of wormholes become essential components of pioneering endeavors. As pioneers chart the course for future generations, they leave an indelible mark on the cosmic landscape, paving the way for unprecedented discoveries and interstellar adventures.

The Uncharted Realms: Humanity's Exploration of the Unknown

In the unexplored realms of the cosmos, humans venture into the unknown, seeking to unveil the secrets that lie beyond the veil of ignorance. This section explores the uncharted territories of the universe, delving into the cosmic mysteries that await discovery. What phenomena and anomalies remain unexplained in the depths of space? How do space colonies prepare for the challenges posed by the unknown? Space colonies, with their astrophysicists and cosmologists, embark on cosmic odysseys, mapping uncharted star systems, investigating cosmic anomalies, and unraveling the enigmas of the multiverse.

The uncharted realms of the universe hold tantalizing secrets, from dark matter and dark energy to parallel universes and cosmic voids. Spacefaring civilizations, equipped with advanced telescopes, quantum sensors, and cosmic mapping technologies, explore uncharted galaxies, studying their composition, gravitational interactions, and cosmic evolution. The study of cosmic voids, the search for gravitational waves, and the investigation of cosmic background radiation become avenues of exploration. As humans probe the unknown, they confront the mysteries of the universe, peering into the cosmic abyss with a sense of awe and anticipation.

The Infinite Frontiers: Humanity's Reach into Infinity

In the infinite expanse of the cosmos, humans extend their reach into infinity, pushing the boundaries of knowledge and understanding. This section contemplates the infinite frontiers that beckon humanity, exploring the cosmic wonders that lie beyond the observable universe. What lies at the edge of the cosmic horizon? How do spacefaring civilizations prepare for the challenges posed by infinite frontiers? Space colonies, with their cosmographers and quantum astronomers, engage in theoretical explorations, contemplating the nature of infinity, cosmic inflation, and the existence of parallel universes.

The infinite frontiers of the universe evoke profound questions about the nature of reality, the fabric of spacetime, and the possibilities of higher dimensions. Spacefaring civilizations, through their particle accelerators, quantum computers, and cosmic observatories, delve into the depths of theoretical physics, exploring concepts such as string theory, brane cosmology, and cosmic strings. The study of higher-dimensional spaces, the exploration of quantum fluctuations in the cosmic microwave background, and the investigation of the cosmic web become avenues of inquiry. As humans peer into the infinite, they confront the mysteries of existence, embarking on a cosmic odyssey that transcends the limits of human perception.

The Endless Horizon: Humanity's Quest for Understanding

In the endless expanse of the cosmos, humanity's quest for understanding stretches towards the horizon, reaching for the stars and beyond. This section reflects on the enduring human spirit that propels space exploration, contemplating the limitless possibilities that await in the unexplored realms of the universe. What drives humans to seek knowledge and understanding in the cosmic abyss? How does the pursuit of understanding shape the destiny of space colonies? Space colonies, with their scholars and cosmic philosophers, engage in contemplative inquiries, exploring the profound questions that define the human condition and the cosmic reality.

The endless horizon of the universe becomes a canvas upon which humans paint their aspirations, dreams, and intellectual pursuits. Spacefaring civilizations, through their scientific endeavors, artistic expressions, and spiritual explorations, strive to comprehend the vastness of the cosmos, embracing the unknown with courage and curiosity. The study of cosmic evolution, the exploration of extraterrestrial life, and the contemplation of the origin of the universe become avenues of enlightenment. As humans gaze upon the endless horizon, they embark on a transformative journey of self-discovery and cosmic understanding, transcending the limitations of mortality and delving into the infinite depths of the cosmic unknown.

Conclusion: The Cosmic Tapestry Unfurls

In the closing chapter of this cosmic odyssey, the threads of the cosmic tapestry converge, weaving a narrative of humanity's exploration, discovery, and enlightenment. The eternal quest for knowledge, the pioneering spirit that propels space exploration, the uncharted territories of the universe, the infinite frontiers that beckon humanity, and the endless horizon of understanding—all these elements form the intricate patterns of the cosmic tapestry. As space colonies and their inhabitants journey through the cosmic expanse, they leave a legacy of exploration and understanding, shaping the destiny of future generations and illuminating the path towards cosmic enlightenment.

In the grand cosmic tapestry, space colonies stand as beacons of human achievement, testaments to the indomitable spirit of exploration and the insatiable thirst for knowledge. They become cosmic hubs of intellectual inquiry, artistic expression, and spiritual contemplation, fostering a harmonious coexistence with the universe. As spacefaring civilizations continue their cosmic odyssey, they carry with them the lessons learned, the wisdom gained, and the profound insights that arise from the exploration of the cosmos.

The cosmic tapestry unfurls, revealing the interconnectedness of all things, the unity amidst diversity, and the boundless potential of the human spirit. In the vast cosmic tableau, space colonies become integral threads, weaving humanity's story into the fabric of the universe. As the cosmic journey continues, propelled by the eternal quest for understanding, humanity's legacy in the cosmos becomes a beacon of enlightenment, inspiring generations to come and illuminating the cosmic tapestry with the brilliance of human ingenuity and curiosity.

Chapter 60: Space Colonies and Interstellar Diplomacy

The Galactic Congress: Uniting Star-Spanning Civilizations

In the vast cosmic arena, the concept of interstellar diplomacy emerges as a beacon of hope, transcending planetary boundaries and fostering unity among diverse spacefaring civilizations. This section delves into the intricate web of diplomatic interactions between star-spanning societies, exploring the treaties, alliances, and peaceful coexistence agreements that form the foundation of interstellar diplomacy. How do spacefaring civilizations navigate the complexities of diplomatic relations in the cosmic expanse? What role do space colonies play in shaping interstellar diplomacy? In the Galactic Congress, representatives from myriad worlds convene to discuss matters of galactic importance, striving for mutual respect and understanding among the stars.

Diplomatic Protocols: Bridging Cultural Chasms

The diplomacy of the stars adheres to a set of protocols designed to bridge the vast cultural chasms that exist among spacefaring civilizations. This section delves into the intricate framework of diplomatic protocols, exploring the nuanced etiquettes, gestures, and languages used in interstellar negotiations. How do diplomats from different civilizations communicate effectively? What cultural sensitivities must be observed to ensure peaceful dialogue? Diplomatic envoys, skilled in the art of cultural diplomacy, navigate the subtleties of language and tradition, fostering an atmosphere of mutual respect and cooperation. As they exchange ideas and knowledge, they weave a tapestry of understanding that transcends interstellar distances, promoting harmony and collaboration among the stars.

The Galactic Accords: Treaties and Alliances in the Cosmic Tapestry

The Galactic Accords stand as a testament to the collaborative efforts of spacefaring civilizations, encapsulating the treaties, alliances, and agreements that define their relations. This section explores the intricacies of the Galactic Accords, from trade agreements and defense pacts to scientific collaborations and cultural exchanges. How do these accords contribute to the stability and prosperity of galactic societies? What role do space colonies play in negotiating and upholding these agreements? Diplomats and legal experts from diverse civilizations converge to draft and ratify the Galactic Accords, fostering cooperation and shared prosperity among star-spanning worlds. Through these agreements, civilizations find common ground, ensuring the peaceful coexistence of species and the flourishing of interstellar trade and knowledge exchange.

The Challenges of Cosmic Diplomacy: Navigating Interstellar Disputes

While interstellar diplomacy promises unity and collaboration, it is not without its challenges. This section delves into the complexities of cosmic diplomacy, exploring the disputes, conflicts, and cultural misunderstandings that sometimes arise among spacefaring civilizations. How do diplomats navigate interstellar conflicts? What mechanisms are in place to resolve disputes and prevent interstellar wars? Mediators and conflict resolution specialists from various worlds employ diplomatic finesse and strategic negotiation, addressing grievances and fostering reconciliation. Through dialogue and compromise, interstellar disputes are resolved, ensuring the continued peace and stability of the galactic community.

The Peaceful Cosmos: Lessons from Intergalactic Alliances

In the tapestry of interstellar diplomacy, there exist shining examples of successful alliances and peaceful coexistence among galaxies. This section examines the intergalactic alliances that have stood the test of time, exploring the lessons learned from these harmonious cosmic unions.

How do alliances contribute to the prosperity and security of entire galactic clusters? What principles of mutual respect and understanding underpin these enduring intergalactic relationships? Ambassadors and historians from galactic alliances share their insights, highlighting the importance of empathy, cooperation, and open dialogue. Through their experiences, they offer valuable lessons for spacefaring civilizations, paving the way for a harmonious and peaceful cosmos where diplomacy triumphs over conflict.

Conclusion: The Cosmic Federation—A Vision of Unity

As this exploration of interstellar diplomacy concludes, a vision emerges—a vision of the Cosmic Federation, a collective of galaxies bound by mutual respect, understanding, and cooperation. Representatives from star-spanning civilizations, including space colonies, gather in the spirit of unity, working towards a harmonious cosmic future. The Cosmic Federation embodies the ideals of peace, collaboration, and the celebration of diversity, fostering a cosmos where diplomatic relations flourish and the collective wisdom of galaxies illuminates the cosmic tapestry. As the Cosmic Federation takes shape, it stands as a testament to the enduring spirit of interstellar diplomacy, guiding the way towards a future where the stars are united in their shared pursuit of harmony and enlightenment.

Chapter 61: Space Colonies and the Unifying Cosmos

In the vast cosmic tapestry, the concept of a unifying cosmos transcends individual stars and galaxies. This chapter explores the profound interconnectedness of all things in the universe, delving into scientific theories that illuminate the fundamental unity underlying the diversity of cosmic phenomena.

The Interconnected Web of Quantum Entanglement

In quantum physics, particles become entangled, instantaneously influencing each other's states regardless of distance. This phenomenon challenges our understanding of reality and hints at a deep cosmic connection. Quantum entanglement, a phenomenon defying classical intuitions, underscores the intricate web binding particles together, suggesting an underlying unity in the fabric of spacetime.

The Holographic Principle: A Glimpse into Cosmic Unity

The holographic principle posits that our three-dimensional reality might be a projection from a two-dimensional surface. This profound idea blurs the lines between dimensions, offering insights into the underlying unity of the cosmos. Examining this principle reveals a fascinating perspective on existence, where information becomes the essence of reality, echoing ancient philosophical concepts of a universal interconnected consciousness.

Cosmic Symmetry: Patterns Across the Universe

Nature exhibits symmetrical patterns, from galaxies' spirals to the fractal structures of cosmic dust clouds. These patterns reflect the universe's inherent order. Exploring cosmic symmetry reveals the mathematical elegance woven into the fabric of reality. From the Fibonacci sequence in spiral galaxies to the Mandelbrot set's self-replicating intricacies, these patterns underscore the interconnected nature of cosmic phenomena.

Entwined Fates: Cosmic Symbiosis of Stars, Planets, and Life

Life's emergence and evolution are intricately linked to cosmic processes. Stars' life cycles influence planets' formation and, in turn, shape the conditions for life. This symbiotic relationship between celestial bodies fosters the conditions for life's diverse manifestations. From extremophiles thriving in extreme environments to complex ecosystems nurturing biodiversity, life's evolution intertwines with cosmic events, exemplifying the unity of existence.

Consciousness and the Cosmic Connection

Beyond physics, the exploration of consciousness and its potential connection to the cosmos becomes a profound inquiry. Ancient wisdom and modern neuroscience converge, suggesting consciousness as a fundamental aspect of the universe. Contemplating altered states, from meditation-induced insights to psychedelic experiences, humanity explores consciousness's expansive potential, hinting at a deeper cosmic connection. Integrating these perspectives fosters a holistic understanding of existence, emphasizing the interdependence of consciousness and the cosmos.

In the contemplation of the unifying cosmos, humanity finds not only scientific enlightenment but also a profound sense of wonder and humility. As we explore the threads binding the cosmos together, we are reminded of our place in the vast interconnectedness of existence, sparking a perpetual quest for understanding and harmony within the unifying fabric of the universe. This chapter stands as a testament to the enduring human spirit, ceaselessly reaching for the cosmic truths that unite all things.

Chapter 62: Space Colonies and the Eternal Wonder

In the boundless expanse of the cosmos, there exists an eternal wonder that transcends the limitations of human comprehension. This chapter delves into the depths of this awe-inspiring fascination, exploring the far-reaching impact of space exploration on the human spirit. From the grandeur of distant galaxies to the intricacies of subatomic particles, the universe evokes profound curiosity and wonder, driving humanity's insatiable thirst for knowledge and discovery.

Galaxies: Cosmic Masterpieces of Wonder

At the heart of the universe lie galaxies, vast conglomerations of stars, gas, and dust. Their breathtaking diversity, from spirals adorned with arms of brilliant stars to ellipticals exuding serene luminosity, captivates the imagination. Galaxies, the building blocks of the cosmos, embody the eternal wonder of cosmic artistry. Contemplating their formations, interactions, and mysteries deepens our appreciation for the vastness of space.

Stellar Nurseries and the Birth of Stars

Within interstellar clouds, stellar nurseries cradle the birth of stars. The process of stellar formation, from collapsing gas clouds to the ignition of nuclear fusion, paints a mesmerizing portrait of the cosmos's creative prowess. Witnessing the birth of stars illuminates the eternal wonder of cosmic genesis, where matter transforms into radiant luminosity, illuminating the cosmic tapestry.

Exoplanets: Worlds Beyond Our Imagination

The discovery of exoplanets, distant worlds orbiting other stars, expands our understanding of the cosmos's diversity. From scorching hot Jupiters to Earth-like habitable zones, each exoplanet unveils the vast possibilities of planetary formations. Exploring these alien realms ignites the eternal wonder of cosmic exploration, sparking contemplation about the potential for extraterrestrial life and the boundless landscapes of the universe.

Quantum Marvels: Wonders at the Subatomic Scale

Delving into the realm of quantum physics, the fabric of reality unravels into a tapestry of mind-boggling phenomena. From entangled particles communicating instantaneously to the probabilistic nature of quantum states, the subatomic world defies conventional intuition. Quantum marvels, the mysteries that underpin reality, evoke eternal wonder, challenging humanity to fathom the profound nature of existence itself.

Microcosmic Mysteries: The Marvels of Life's Building Blocks

Life's foundation resides in the intricacies of molecules and cells, where the dance of atoms shapes the tapestry of existence. Exploring the microscopic world, from DNA's elegant structure to the intricate processes within living cells, illuminates the complexity of life. Microcosmic mysteries, the marvels of life's building blocks, evoke wonder at the intricacies of biological existence, underscoring the eternal fascination with life's origins.

Cosmic Perspective: Nurturing Wonder in Space Colonies

Within the controlled environments of space colonies, humanity gains a unique vantage point. The cosmic perspective, viewing Earth as a fragile oasis in the cosmic sea, instills a profound sense of wonder and humility. Observing celestial phenomena, conducting experiments in microgravity, and contemplating the universe from the colony's confines nurtures the eternal wonder that defines human exploration. Space colonies become crucibles of curiosity, fostering an enduring fascination with the cosmos.

Inspiration from the Stars: Art, Literature, and Music

Throughout history, the cosmos has inspired art, literature, and music, becoming a muse for human creativity. From Van Gogh's Starry Night to the cosmic odysseys of science fiction novels, the wonders of space kindle the flames of artistic expression. Celestial melodies and cosmic poetry echo the eternal wonder that permeates the universe. Examining these creative expressions reveals the profound impact of space exploration on humanity's artistic soul.

Cosmic Curiosity: Humanity's Guiding Light

At the heart of humanity's quest for knowledge lies cosmic curiosity—an insatiable desire to understand the universe's mysteries. This eternal wonder serves as the guiding light, propelling scientists, artists, and dreamers alike to explore the unknown. From the depths of Earth's oceans to the farthest reaches of space, human curiosity knows no bounds. Cosmic curiosity becomes the engine of exploration, inspiring future generations to continue the eternal quest for understanding, knowledge, and the boundless wonders of the cosmos.

Chapter 63: Space Colonies and the Future Beyond

In the vast expanse of the cosmos, the future unfolds with infinite possibilities, beckoning humanity towards uncharted realms of discovery and innovation. This chapter delves into the boundless potential that lies beyond the pages of this book, exploring the ongoing exploration, groundbreaking discoveries, and the pivotal role that future generations will play in shaping the destiny of spacefaring civilizations.

Continued Exploration: Journeying Deeper into the Cosmic Unknown

As spacefaring technologies advance, humanity's reach extends further into the cosmic unknown. The exploration of distant planets, asteroids, and interstellar space becomes an ever-expanding endeavor. Robotic probes, sophisticated telescopes, and interstellar missions delve into unexplored territories, unveiling the secrets of alien worlds and unraveling the mysteries of the universe's distant corners.

Innovations in Propulsion: Bridging the Vast Cosmic Distances

The future of space exploration hinges on revolutionary propulsion technologies. Concepts like ion drives, nuclear propulsion, and antimatter engines hold the promise of bridging vast cosmic distances. By harnessing the power of exotic fuels and cutting-edge engineering, humanity envisions interstellar travel becoming a reality, allowing for expeditions to neighboring star systems and perhaps even beyond the boundaries of our galaxy.

Interstellar Communication: Bridging the Silence of Cosmic Distances

Communication across interstellar distances poses unique challenges. Future generations will grapple with developing advanced communication protocols that surmount the limitations of light-speed. Quantum entanglement, subspace communication, and other speculative technologies might hold the key to bridging the silence of cosmic distances, enabling real-time exchanges between distant star systems.

Harvesting Resources Beyond Earth: The Age of Extraterrestrial Mining

In the quest for resource sustainability, humanity looks towards celestial bodies for abundant raw materials. Asteroid mining, lunar excavation, and the utilization of extraterrestrial resources become imperative for the expansion of space colonies. The age of extraterrestrial mining dawns, heralding a new era where the vast wealth of the cosmos becomes accessible, ensuring the survival and growth of spacefaring civilizations.

Colonization of Exoplanets: Seeding Life Among the Stars

The discovery of habitable exoplanets fuels the dream of interstellar colonization. Future generations may embark on daring missions to these distant worlds, seeding life among the stars. Terraforming, genetic adaptation, and ecological engineering become the tools of cosmic pioneers, transforming alien environments into havens for human habitation, establishing outposts that serve as stepping stones towards the colonization of the galaxy.

Ethical Considerations and Cosmic Stewardship: Navigating the Moral Landscape

With the power to shape entire worlds comes profound ethical responsibility. Future spacefarers must grapple with ethical dilemmas surrounding planetary terraforming, the preservation of alien ecosystems, and interactions with potential extraterrestrial life. Cosmic stewardship becomes paramount, demanding careful navigation of the moral landscape to ensure the harmony and coexistence of diverse life forms in the vastness of space.

The Role of Artificial Intelligence: Collaborative Intelligence in Space Exploration

Artificial intelligence (AI) becomes an indispensable partner in the cosmic journey. Advanced AI systems aid in interstellar navigation, complex simulations, and decision-making processes. Collaborative intelligence between humans and AI enhances the efficiency and safety of space exploration, ushering in an era where human ingenuity and machine learning work in tandem to unravel the cosmic mysteries.

The Infinite Tapestry of Possibilities: The Future Written in the Stars

In the tapestry of the cosmos, humanity discovers the infinite threads of possibilities waiting to be woven into the fabric of the future. The collaboration between diverse cultures, the pursuit of scientific inquiry, and the boundless creativity of human imagination paint a vibrant portrait of what lies ahead. The future, written in the stars, is a testament to the enduring spirit of exploration that drives humanity towards a destiny among the celestial realms.

As this chapter concludes, it leaves the reader with a profound sense of anticipation. The cosmic frontier, once a realm of distant dreams, is now within reach.

Chapter 64: Space Colonies and the Reader's Role

In the final chapter of this cosmic odyssey, we turn our attention to you, the reader, who has embarked on this unparalleled journey through the realms of space and imagination. Within these pages, you have ventured across the cosmic expanse, delving into the mysteries of the universe and contemplating the boundless possibilities that await in the farthest reaches of the cosmos. This chapter serves as a tribute to your insatiable curiosity, your willingness to explore the unknown, and your role in the grand tapestry of the cosmic narrative.

A Journey Shared: The Bond Between Reader and Cosmic Explorer

As a reader, you have become a cosmic explorer, traversing the galaxies and witnessing the birth and death of stars. Through the words on these pages, you have transcended the confines of Earth, embracing the vastness of space and time. The bond between the reader and the cosmic explorer is a testament to the power of storytelling, connecting minds across the universe and illuminating the wonders of the cosmos.

The Curiosity Within: Nurturing the Spirit of Inquiry

Your curiosity is a beacon guiding you through the labyrinthine wonders of the universe. It is the driving force that propels you to ask questions, seek answers, and venture into the unknown. The spirit of inquiry within you is a flame that must be nurtured, for it is through curiosity that humanity has made its most profound discoveries, unraveling the secrets of the cosmos one question at a time.

Imagination Unbound: Fueling the Cosmic Odyssey

Imagination is the vessel through which you have sailed the celestial seas, exploring distant planets, envisioning advanced civilizations, and pondering the existence of extraterrestrial life. It is the creative force that fuels the cosmic odyssey, allowing you to dream of futures yet unwritten and realities yet undiscovered. Your imagination, boundless and limitless, serves as the engine propelling you towards new horizons of understanding and wonder.

The Infinite Possibilities: Your Role in Shaping the Cosmic Narrative

Within the vast expanse of the universe, you are not merely an observer but an active participant in the unfolding cosmic narrative. Your thoughts, ideas, and aspirations shape the way you perceive the cosmos, influencing the stories you tell and the questions you ask. The infinite possibilities that lie before you are a canvas upon which you can paint your own cosmic adventures, contributing unique hues to the ever-expanding tapestry of the universe.

Gratitude for the Journey: Acknowledging the Reader's Contribution

In the cosmic dance of existence, your presence as a reader is a cherished contribution to the collective human understanding of the universe. Your engagement with these words, your contemplation of cosmic concepts, and your willingness to explore the unknown have added depth and richness to the narrative. The authors, scientists, and storytellers who have crafted this book extend their heartfelt gratitude for the role you have played in this cosmic journey.

Continuing the Exploration: Embracing the Infinite Horizons Ahead

As you reach the final pages of this book, remember that the cosmic odyssey does not end here; it merely transforms. The knowledge you have gained, the wonders you have witnessed, and the questions you have pondered are stepping stones towards a future filled with endless exploration. Embrace the infinite horizons ahead, for the universe, with all its enigmas and marvels, awaits your further discovery.

A Call to Action: Inspiring Future Explorers

In closing, this chapter stands as a call to action, inspiring not only you, the reader, but future generations of cosmic explorers. Encourage others to delve into the mysteries of the universe, to question the nature of reality, and to dream of interstellar adventures. Your enthusiasm and passion for the cosmos can ignite the spark of curiosity in others, ensuring that the spirit of exploration continues to illuminate the human journey for generations to come.

A Universe of Possibilities: Farewell, But Not Goodbye

As we bid farewell to the pages of this book, remember that the universe is vast, diverse, and ever-changing. Farewell, but not goodbye, for the cosmic wonders and the unexplored territories of the cosmos will always beckon. The story of space colonies and the human spirit's cosmic odyssey continues, and you, dear reader, are an integral part of this ongoing saga.

In the grand tapestry of the universe, your role as a reader is celebrated and cherished. May your future explorations be filled with awe, discovery, and the unwavering belief in the limitless potential of humanity. As you close this final chapter, remember that the cosmic adventure continues, and the universe, with all its marvels and mysteries, eagerly awaits your return.

Conclusion:

Charting Our Cosmic Legacy and the Evolution of Humanity

In the boundless expanse of the cosmos, humanity has embarked on a remarkable journey, reaching out to the stars and establishing colonies beyond the confines of Earth. Through the pages of this book, we have delved deep into the intricate tapestry of space colonization, exploring the profound impact it has on our understanding of the universe and the very essence of what it means to be human.

From the pioneering spirit of the first colonists to the awe-inspiring scientific discoveries within these cosmic habitats, our exploration of space has reshaped our perspective, challenging our boundaries and inspiring us to dream beyond the stars. We have witnessed the resilience of the human spirit, adapting to the challenges of space and embracing the unknown with courage and curiosity.

As we reflect on the pages turned and the knowledge gained, it becomes evident that space colonies represent more than just a new frontier. They stand as a testament to human ingenuity, cooperation, and our unyielding desire to explore the unknown. Within these colonies, we have not only extended our reach into the cosmos but have also found new ways to collaborate, innovate, and harmonize with the cosmic rhythms.

The legacy of space colonization echoes with lessons carved in the cosmos—lessons of sustainable living, interdisciplinary collaboration, and the unbreakable spirit of exploration. It challenges us to think beyond the confines of our planet, fostering a sense of unity and shared destiny that transcends borders and cultures. The cosmic journey, as charted within these pages, is not just a vision of our future—it is a testament to our potential, a call to embrace our role as stewards of the universe.

As we bid farewell to these written words, let us carry the spirit of exploration forward. Let us continue to gaze at the stars with wonder, to question the mysteries of the universe, and to nurture the curiosity that propels us toward new horizons. Together, we are charting our cosmic legacy, shaping the future not just for ourselves but for the generations yet to come. In the vastness of space, our story continues—a story of boundless possibilities, endless curiosity, and the eternal quest for understanding.